BOLTS from the BLUE

Also by Ralph Herman Abraham

• Foundations of Mechanics (with Jerrold Marsden)

• Manifolds, Tensors, and Applications
 (with Jerrold Marsden and Tudor Ratiu)

• Dynamics, the Geometry of Behavior
 (with Christopher Shaw)

• Chaos, Gaia, Eros

• Chaos in Discrete Dynamical Systems
 (with Laura Gardini and Christian Mira)

• Chaos, Creativity, and Cosmic Consciousness
 (with Terence McKenna and Rupert Sheldrake)

• The Evolutionary Mind
 (with Terence McKenna and Rupert Sheldrake)

• The Chaos Avant-garde (with Yoshisuke Ueda)

• Demystifying the Akasha (with Sisir Roy)

BOLTS from the BLUE

Art, Mathematics, and Cultural Evolution

Ralph Herman Abraham

Epigraph Books
Rhinebeck, New York

Bolts from the Blue: Art, Mathematics, and Cultural Evolution
© 2011 by Ralph Abraham.

Printed in the United States of America

Epigraph Books
27 Lamoree Road
Rhinebeck, New York 12572
www.epigraphPS.com
USA 845-876-4861

ISBN 978-0-9830517-7-0

Library of Congress Control Number: 2010943493

Bulk purchase discounts for educational or promotional purposes
are available. Contact the publisher for more information.

Dedicated to the late:

Andra Akers

Terence McKenna

Nina Graboi

FOREWORD
by William Irwin Thompson

The End of the Age of Religion and the Birth of Symbiotic Consciousness[1]

Through my collaboration with the chaos mathematician Ralph Abraham in designing an evolution of consciousness curriculum for the Ross School in East Hampton, New York, I began to understand that the shift from the linear causation of Galilean dynamics in the early modern era to the complex dynamical systems of our era also expressed a shift from linear modernist ideologies and religions to planetary ecologies of consciousness in which diversity was affirmed. In the evolution of the catastrophe theory of the 1960s—with their images of saddles and butterfly folds—to the images of fractals and Lorenz attractors in the chaos dynamics of the 1980s, our cultural Imaginary was given a gift of a new alphabet of symbols. Dynamical systems were given geometrical portraits of their behavior, and these were therefore called phase portraits.[2] The linearity of left-brain thinking was now to be balanced with a right-brain activation. This emergence of a new visual mathematics expressed, in effect, a return on a higher turn of the spiral to hieroglyphic thinking.

It all started with Poincaré in Paris in 1889 when he showed that the clean and consistent system of Kepler in which the planets rotated around the sun in neat ellipses was not correct, that the solar system was actually a chaotic system. You can date the birth of complex dynamical systems with Poincaré and say that the new era begins with his mathematical revisioning of the geometry of behavior of the solar system. At about this time the premodernist esoteric cosmologies began to experience what Marshall McLuhan called "cultural retrieval," and thinkers like Rudolf Steiner, Hazrat Inayat Khan and William Butler Yeats began their visionary careers. The linear reductionism of modernism was go-

---

1   Reprinted from *Seven Pillars House of Wisdom*.
2   (Abraham, 1999)

ing to be challenged by a cultural retrieval of animism on one side and higher mathematics on the other. The composer Satie was a Rosicrucian, and the painters Kandinsky and Mondrian were Theosophists. Clearly, complex dynamical systems began to impact on the cultural evolution of human spirituality.

What could this new planetary culture possibly look like? First, egocentric monumentality and the extensive clutter of industrial civilization could be eliminated. We could shift from industrial object to ecological process—as foreshadowed in the "Living Machines" of John Todd.[3] Some buildings through the effectiveness of nanotechnologies could become ephemeral and evanescent; enduring structures could be more ecologically embedded in their setting—like the Shire of the Hobbits in the writings of Tolkien. We could become electronic nomads who pitch their tent, and then pack up and move on. Buildings could become appliances that we turn on with a switch, and then turn off to make them disappear in the forest or meadow, and this would enable the human and animal domains to coexist more peaceably. Think of this as an electronic version of the Arthurian Lady of the Lake who used enchantment to keep her settlement hidden to mortals so that it appeared to the local inhabitants only as a lake.

Our machines could become intimate and ensouled by the elemental beings whose presence we might rediscover in the coming period of intense volcanic and tectonic activity. We will all have a chance to become animists again—like the present population of Iceland or the kahunas of Polynesia—those people who have been living with volcanoes for some time. The kind of being that once was envisioned as ensouling a sacred mountain could now be seen to ensoul the noetic lattice of crystals in electronic and quantum computers in a new cultural Imaginary. As these computers are worn on and in our bodies and our body-politic, our sense of "in here" and "out there" would be transformed as a cube became a tesseract or a sphere became a spiraling hypertorus in which the inside and outside surface are continuous through the spiraling axis.

Our consciousness could become symbiotic, as elementals become to us what mitochondria are to nucleated cells. But this symbiotic consciousness need not simply be restricted to human and elemental or

---

3   (Todd, 1994)

animal realms, it could also be extended to involve the celestial intelligences. To imagine these "software" beings that are made out of music and mathematics, we need once again to go back to the end of the film 2001: A Space Odyssey (which I discussed with Arthur C. Clarke over breakfast in New York long ago in 1971). When the astronaut approaches the monolith in orbit over Jupiter, he sees coming toward him rotating crystals of light and complex pulsing topologies. These are Stanley Kubrick's technological envisioning of what esoteric initiates would recognize to be the Neoplatonic "celestial intelligences"—the Jinn of the Moon and the Angels and Archangels of the planets and stars. For Kubrick's and Clarke's vision, however, it is these high tech cosmic beings who serve as midwives to the astronaut's rebirth as he moves out of history into myth.

To draw a circle one moves from the point to the line; to draw a sphere, one pulls the circle up into the third dimension; to create a hypersphere, one rotates the sphere into the fourth dimension. Our physical body, or what the yogis call our food sheath, has three dimensions, but our other bodies or sheaths have more dimensions, and it is in the facets of the topology of these dimensions that the celestial intelligences can interface with us and participate in the field of our consciousness. The brain may be a three-dimensional volume, but neurons in separate parts of the brain can fire together in the neuronal synchrony of the range of 40 Hertz. The geometry of the synchronies engage as facets of the higher-dimensional geometries of the subtle bodies—where both the Dalai Lama and Rudolf Steiner say memory is stored—so the play of consciousness should not simply be reduced to a section of the brain.[4] Cultures have in the past called this process of consciousness, imagination or intuition, but whatever one calls it, it is basic to the creative process in art, science and spiritual contemplative practice.

Mediating between the elemental and human on one polarity and the celestial intelligences on the other is the realm of the soul, which for men is often figured as a feminine being—a Tara or Beatrice. (Jung maintained that for women this contrasexual "animus" was male—the Christ of Saint Teresa of Avila or the Krishna of Mirabai.) This being appears regularly in the intermediate life of our dreams—and here one needs to understand that dreams, as Sri Aurobindo pointed out for his

---

4 (Varela, 1997; p. 174)

generation, are very muddled memories of higher spiritual experiences blended with the proprioperceptions of the physical body and the brain's return to waking consciousness. As the spirit returns to the confinements of incarnation, it can start to dream it is in a conference, or a crowded airport, and as it becomes aware of the body's full bladder, it will begin to dream that it is looking for the restroom in the airport. To interpret these dreams with Freudian or Jungian symbolic systems is, at this level, a category mistake. The imagery is taken from personal memory and is being used as metaphors for the reactivation of specific brain modules that are operative in the cognitive functions of the waking mind.

Shamanism was the form of spirituality that evolved in the oral culture of preliterate societies. Religion was the form of spirituality that emerged with literate societies and their new temple-based readings of the stars and sacred texts. Though traditionalists may wish time to stop, it does go on, and now in our global electronic society, a new transreligious form of spirituality is emerging, one that will not replace religions, anymore than the nucleus of the cell replaced the mitochondria, but will envelop them in a much vaster form of consciousness. In this futuristic ontology, we are already beginning to glimpse an evolutionary Entelechy—a symbiotic consciousness of human, elemental, psychic and celestial intelligences. In the smuggled esoterism of children's literature, comic books and science fiction, an archetypal group of four becoming one is being foreshadowed. For example, we see this grouping expressed in Fantastic Four and The Wizard of Oz.

In preindustrial animist cultures, the human would establish a place for the elementals to cohabit or participate in its human life through the intermediary use of a magical object—a magical ring or stone, an Aladdin's lamp or an ensouled sword such as Roland's Durendal or Charlemagne's Montjoie. Like the needles used in acupuncture, this numinous object can interact with its possessor at the subtle-physical or etheric level—the level of qi or prana. In shamanistic cultures, the individual would project out of his or her body and travel in a spiritual or astral world. If one is able through meditation to remain watchfully awake in the state of deep dreamless sleep, then one experiences a vast magenta sea of cognitive bliss in which one hears the music of everything—every existent being in the universe sounding its presence. This is the world of the Holy Spirit that is "above" or "below"—remember these terms express a merely Euclidean geometry—the psychic realm of dreams or astral out-of-the-body travel. In the astral, one travels, but in the realm of celestial music, the center is everywhere and the circumference nowhere, so there is no need to move. One simply joins in this Hallelujah Chorus to the nth power by listening and sounding one's ontic note. In this sense, music is not a representational art of mimesis, but an ontological performance.

The human is the ordinary ego in time, but in the completion of our emergent evolutionary spiritual process the ego becomes transhuman, "anointed" or Christic. For esoteric Christians, the prophet Jesus became the Christ at the time of the baptism by John, and this narrative of the Son of Man describes the process of enlightened individuation. For Buddhists, this process is seen more like a wave than a particle, one in which ego-hood is transformed in dependent co-origination (pratityasamutpadha) of enlightened Buddha Mind.[5]

Now you may have noticed that one thing that results from this ontology of symbiotic consciousness is a non-locality in which "out there" is "in here" so that it is no longer necessary to put three-dimensional bodies in expensive tin cans and space suits and try to propel them to the stars. We may be able to go to the Moon, and my Lindisfarne colleague James Lovelock's proposed atmospheric and bacterial "greening of Mars" would certainly be a worthy project for sublimating the nations' defense industries into a transnational technological project with Europe, Russia, Japan and China, but I think it is highly unlikely humans could travel in physical bodies to the stars. Indeed, that is precisely what the astronaut of 2001 discovers in his transformation from technological man to star child. So the Christian fundamentalist notion that we can trash the Earth and move on, and that whatever mess we make here is permitted because Jesus will play the role of a suburban mom coming in at the end of the day to clean up our room for us and then take us away in a vacation rapture to some theme park heaven is an expression of folk superstition and the limited three-dimensional thinking of religion.

Jean Gebser taught us that when a new evolutionary form becomes efficient, the old becomes deficient.[6] When religion emerged, shaman-

---

5 (Varela, 1997; p. 174)

6 (Gebser, 1984; p. 93ff)

ism decayed into sorcery and black magic. Now that a new planetary spirituality is emerging, religion has become a toxic dump, as witnessed in the recent terrorist attacks in Mumbai.

But the religionists are right in one way; it is the end of their world, but that also means the end of the age of religion and the beginning of a unique/universal self-similar architecture of consciousness that is based upon individual experience and not upon priestcraft, rigid dogma and collective forceful indoctrination. If we can avoid the dark age of religion that now stares us in the face, we may discover a more surprising and delightful politics of Being behind the mask.

PREFACE

Andra Akers was the stimulus and cheerleader for much revolutionary thought during the 1980s and 1990s in Los Angeles and New York until her premature death, and we miss her. William Irwin Thompson and I have collaborated in sharing ideas since our meeting in 1985 at Andra Akers' home in West Hollywood.

In the context of our cooperation over these years, our particular ways of looking at the development of art and mathematics in world cultural history converged into a narrative with a mathematical frame. This book owes a huge debt to Bill Thompson, whose ideas sparked four of the six chapters, and added much to the glue and integrity of the whole project.

For crucial editorial help and all kinds of advice and encouragement over four decades, I am especially grateful to my friend, philosopher Paul A. Lee.

For keen and useful feedback on this book, many thanks to Claudia l'Amoreaux, and Kevin Cashen. For Chapter 1, thanks to Ernest Mc-Clain and Edmund Carpenter for their comments on an early draft. And for publication assistance, many thanks to Paul Cohen and his team at Epigraph Books.

Ralph Herman Abraham
Santa Cruz, California
October 26, 2010

Our title refers to a singular spark of inspiration in the creative synergy of the arts and mathematics. For example:

> *The invention of logarithms came on the world as a bolt from the blue. No previous work had led up to it, foreshadowed it or heralded its arrival. It stands isolated, breaking in upon human thought abruptly without borrowing from the work of other intellects or following known lines of mathematical thought.*[1]

---

1   Inaugural address by Lord Moulton, on the 300th anniversary of the invention of logarithms (a radically new way to multiply numbers) by John Napier, Edinburgh, 1914. Published as "The invention of logarithms," in *Napier Tercentenary Memorial Volume*, p. 3. Also quoted by Eli Maor, in: *The Story of a Number*, Princeton, 1994, p. 13.

CONTENTS

INTRODUCTION

This book is about the role of mathematics in the evolution of culture, and the evolution of mathematics itself. It aims to clarify the question: What is mathematics, and who is a mathematician? We believe that everyone has a native mathematical talent, before it is tainted in school by math anxiety. In this book we give six examples of bolts from the blue, in which an artist and self-taught mathematician brings forth an important new mathematical idea in an intuitive revelation, an innovation that triggers a major transformation within mathematics, and in cultural history as well. The book is organized around three main themes:

• *Dynamical historiography*, that is, world cultural history regarded as a complex dynamical system, a network of cultural ecologies, a history evolving through epochs (plateaus) segmented by *bifurcations* (generalized paradigm shifts). *Chaos theory*—comprising the new computer-based mathematical theories of nonlinear dynamics, fractals, chaos, bifurcations, complexity, neural networks, and so on—provides a new way of looking at any complex dynamical system. The first of our three main themes is based on this new way of looking at one of the largest complex systems of all: world cultural history[1]

• *Mathematical mentality*, that is, the dominant style of mathematical cognition exhibited by individuals in a cultural ecology. We will describe five mentalities from the viewpoint of dynamical historiography, cultural ecology, and bifurcation theory. This theme is a math-centric kind of cognitive psychology, in which we characterize an entire cultural ecology by its mathematical style. This is analogous to art history being divided into Ancient, Medieval, and Modern, for example.

• *Bolts from the blue*, that is, the reception of a new mathematical strategy by an artist or intellectual, as if by telepathy from the stars, leading to a bifurcation in a cultural-historical system.

Our six chapters are exemplary of our third theme: bolts from the

---

1   See (Abraham, 1994) for an early account of this view. One of the first historians to adopt the chaos theoretic view is William Irwin Thompson.

blue. But taken together, they also illustrate a subtheme: *Anyone can be an important mathematician, if not handicapped in school by math anxiety.* This leads us to an urgent problem: how to design a school math curriculum that avoids math anxiety. We now explain these themes, one at a time.

Dynamical historiography

The new mathematics of complex dynamical systems, that we call chaos theory for short, emerging into the mainstream since 1970 or so, has provided a new cognitive style that we call the *chaos dynamical mentality.* The main concepts here, derived from the new mathematical theories, include attractors, basins of attraction, bifurcations, the emergency of new attractors, self organization, pattern formation, and many more. For details of these new concepts one may refer to other books.

But the new view of world cultural history, called dynamical historiography, is fully explored in my book, *Chaos, Gaia, Eros.*[2] The word bifurcation, a technical term from chaos theory, provides more precision to the ideas of paradigm shift (from Ludwik Fleck[3] and Thomas Kuhn[4]) and cultural mutation (of Jean Gebser[5]).

In chaos theory, there are three kinds of bifurcation: catastrophic bifurcation (such as a quantum leap), subtle bifurcation (such as the gradual onset of vibration in a machine), and explosive bifurcation (in which a chaotic state suddenly expands in magnitude, like an earthquake).

The sequence of five stages

In the l960s and 70s, William Irwin Thompson joined anthropology, artistic studies, and political history into a unique approach to cultural

---

2   (Abraham, 1994)

3   Fleck, a holocaust survivor, introduced the paradigm shift idea in his book, Entstehung und Entwicklung einer wissenschafilichen Tatsache; Einfuhrung in die Lehre von Denkstil und Denkkollectiv, of 1935. See the English edn., (Fleck, 1979).

4   Kuhn, inspired by Fleck, based his paradigm theory on the example of the Copernican revolution.(Kuhn, l962)

5   Gebser presented a complete theory of the evolution of consciousness in his book, Ursprung und Gegenwart, of 1966. For the English, see (Gebser, 1984).

history.[6] He intuitively made use of the ideas of dynamical historiography to parse our whole history into plateaus punctuated by major shifts.

In his *Pacific Shift* of 1985, he described four major stages in the history of the West: the Riverine, Mediterranean, Atlantic, and Pacific-Space stages.[7] The approximate dates for the major bifurcations separating them (and their chief characteristic features) are:

- Tl. Beginning of Riverine: 4000 BCE[8] (writing)
- T2. Riverine to Mediterranean: 2000 BCE (alphabet)
- T3. Mediterranean to Atlantic: 1500 CE (printing)
- T4. Atlantic to Pacific-Space: 2000 CE (computer)

The Riverine refers to the cultural ecology of the Indus, Nile, and Mesopotamian valleys. The Mediterranean is the cultural ecology all around the Mediterranean Sea. The Atlantic includes Western Europe and Eastern North America, in particular. And the Pacific cultural ecology is that currently emerging around the Pacific Rim. Hence his title, *Pacific Shift*. We are now in the midst of a major bifurcation (a tipping point) from the Atlantic to the Pacific cultural ecology.

Independently, in my *Chaos, Gaia, and Eros* of 1994, I presented world cultural history in four epochs, divided by three major bifurcations:

- Al. Paleolithic to Neolithic: 10,000 BCE (agriculture)
- A2. History: 4000 BCE (writing)
- A3. Chaos: 2000 CE (computational math)

In subsequent joint work, we settled on the sequence of five epochs (and mathematical mentalities) described below, which frame the structure of this book. Ignoring the Neolithic bifurcation, Al, the Thompson sequence is a refinement of my sequence. A2 coincides with Tl, and A3 with T4. Our joint sequence, includes: Al, A2=Tl, T3, A3=T4, and replaces T2 with a new shift between T1 and T3, the Islamic (see list below).

---

6 See (Thompson, 1967) and (Thompson, 1971).

7 Developed in a talk on February 13, 1983. See (Thompson, 1985, esp. Ch. 4).

8 In this book we use BCE and CE in place of BC and AD for dates before and after the time of the Christ.

The main motivation for our special emphasis on bifurcations is to understand our current transformation. To participate wisely in the creation of the next cultural ecology, we must study the major bifurcations of the past.

Between any two consecutive large bifurcations there may be several medium bifurcations, and between them, many small ones. Two medium bifurcations within the Mediterranean epoch figure in this book:

- the enlightenment of early Islam, 800 CE, and
- the Italian Renaissance, 1400 CE.

To some extent our periodization into five stages is arbitrary. We could have had a longer list, regarding finer structure in the evolution of mathematics and cognitive styles. But we regard our five stage sequence as the shortest that is consistent with the history of mathematics, as normally understood by mathematicians. This is the sequence of periods basic to this book, with very approximate starting dates:

- Prehistoric, from 100,000 BCE
- Historic, 4,000 BCE
- Islamic, 800 CE
- Dynamic, 1400
- Chaotic, 2000

Cultural ecologies

Our concept of cultural ecology is derived from *Gaia theory*. Developed by Earth scientist James Lovelock in a series of talks, articles, and books since 1968, this is a massively integrated theory of the complex ecosystem comprising the Earth's oceans, land masses, biosphere, atmosphere, energy balance with the sun and space, and so on.[9] Building on earlier work by Vernadsky in Russia, Lovelock taught the principles of complex dynamical systems in the context of our living Earth, as he called this massive ecosystem. In this book we are taking Gaia one

---

9   For Gaia theory, see all books by James Lovelock, for example, (Lovelock, 1995). Also, (Abraham, 1994).

step further, to consider world cultural history as an integral part of the big picture.

The concept of cultural ecology is an application of complex dynamics to cultural history, that is, an example of dynamical historiography. Historical and ecological ideas are combined, and applied to world cultural history. Cultural ecology regards a culture as an ecosystem. Its parts—for example, literature, visual arts, musical creations and performance, mathematical developments and applications, scientific discoveries, economics, etc.—are interconnected like the flora, fauna, and environment in a biospheric ecosystem.[10] Local cultural microsystems are interconnected in a global cultural macrosystem, linked by trade, cultural diffusion, and so on. Cultural and biospheric systems are interconnected in Gaian superphysiology. This giant web evolved, and continues to evolve, as a complex dynamical system, with emergent properties, bifurcations, etc.

Each cultural ecology is characterized by attributes, including a dominant mathematical mentality.[11]

Mathematical Mentalities

By mathematical mentality we mean more than specific mathematical knowledge. We mean a characteristic cognitive style in approaching perception, analysis, reasoning, magical operations, and so on. It is characteristic of the whole intellectual approach of a cultural ecology. Here is the list of the five cultural ecologies, their mathematical mentalities, and their mnemonic codes, basic to this book:

- Paleolithic cultural ecology, aRithmetic mentality (R)
- Riverine, Geometric (G)
- Islamic, Algebraic (A)
- Renaissance, Galilean Dynamical (D)
- Modernist, Chaos Dynamical (X)

We use RGADX as a mnemonic for this sequence. Each cultural

---

10 For the science of biospherics originally due to Vernadsky, see (Snyder, 1885).

11 See (Thompson, 2004).

ecology and math mentality is separated from its sequel by a cultural bifurcation. Our chapters will clarify these systems with examples.

How we came to this picture

The sequence of cultures and mentalities, RGADX, is just one of many that might occur to an amateur historian. Why and how did we come to this? We must now confess that it has evolved through extensive efforts to revise the school curriculum so that mathematics and cultural history might be integrated, and math anxiety avoided. We intend that this sequence be used as the outline of a school curriculum in middle and high schools around the world. Used how? This book aims to give an indication of integrated teaching units, taught in sequence, along with the related math skills motivated by the stories.

Bolts from the blue

The six chapters of this book are intended as exemplary, self-standing units, defining our basic concepts, and leading to the full integration of math and cultural history. In addition, each is the story of a bolt from the blue, precipitating a cultural bifurcation. They are presented here in historical order.

The first is devoted to the *Venus of Lespugue*. This is a small paleo-lithic sculpture, one of a number of similar objects found all over Old Europe. It belongs to the paleolithic, prehistoric past, an represents the essence of the aRithmetic Mentality around 23,000 BCE. Its dimensions conform to a precise set of ratios, derived from the musical intervals of the Greek Doric scale. The Venus was created thousands of years before the beginning of the Riverine ecosystem in Mesopotamia, with its Geometric Mentality. Advanced arithmetic was developed by paleolithic musicians, and carved in stone for the future.

The second is devoted to the *Bethels of Scotland*. These are small sculptures the size of a baseball, carved in one of the hardest stones of Europe. Many bethels, dating from the time of the megalithic monuments such as Stonehenge, have been found all over Scotland. Their shapes include the cosmic figures, also known as Platonic solids much later in ancient Greece. They demonstrate the solid geometry of the megalithic phase of the Riverine cultural system, as it approaches the bifurcation from the aRithmetic to the Geometric Mentality. Sublime geometrical knowledge was intuited by megalithic sculptors, and again, saved for posterity in stone, around 3000 BCE.

The third, on the origins of algebra and the Algebraic Mentality, discloses a smaller but thrilling bifurcation within the Mediterranean cultural ecology, in which the aRithmetic Mentality is revived, along with its association with writing, and extended: a Riverine renaissance, preserved in manuscript form.

The fourth, following the revival of perspective by artists of the Renaissance, demonstrates the anticipation by the painter Fra Angelico, in his geometry of angels, of an advanced result in the branch of mathematics called topology, which emerged 300 years later in academic circles in the 20th century, at the beginning of the Pacific Shift. This is an extraordinary case of mathematical clairvoyance, preserved as a painting: an archetypal bolt from the blue.

The fifth, on the birth of the Dynamical Mentality, gives us a step-by-step analysis of a bifurcation, seen from the complex dynamical point-of-view characteristic of our new Chaos Dynamical Mentality.

The sixth and final chapter reveals the mathematical precognition by a modem painter of *fractal geometry*, fundamental to the Chaos Dynamical Mentality, 60 years before fractal images were created for the first time by computer graphics, in the work of Benoit Mandelbrot. A bolt from the blue, again sent forward to us as a painting.

In each case, a *bolt*—an intuitive leap, major advance for mathematics, and trigger of a cultural bifurcation—has been received and recorded from the *blue,* the intuition of an artist or intellectual.

Math anxiety

This book is motivated to show the role of artists and intellectuals in the evolution of mathematics, and to promote Thompson's scheme for world cultural history. But its chief concern is the perilous situation of mathematics in our contemporary society. For there is now a pandemic of math anxiety. And believing as I do that our future is doomed without the intellectual and cognitive support of a healthy and vigorous mathematical culture, this situation demands attention.

After more than fifty years of teaching math in universities in several countries, I have developed the conviction that every person is born with a substantial talent for math which is subsequently destroyed in our schools by a faulty pedagogy that has become traditional during the last century or so. Over the years, I have identified three major flaws. This book is intended to help remedy them.

Flaw #1: No graphics.

The first flaw came to my attention around 1974, when computer graphics first arrived at my university. After creating computer graphic software for research in chaos theory, then a new branch of mathematics, we adapted the hardware and software to support our lower division math courses: calculus, linear algebra, differential equations, and so on. With support from the State of California, these efforts evolved into a major program called the Visual Math Project. Computer graphic illustrations and animations were piped into classrooms using television cables.

We discovered that many students were saved from math anxiety and became successful students of mathematics. Some were so enthusiastic that they became programmers in our project, developing software to teach others what they had learned.

Mathematicians communicate among themselves by coordinating multiple intelligences: verbal, graphic, and symbolic. For example, one cannot learn math without graphics. As our school math programs present math without adequate graphics, learning is handicapped. Students fail to learn, and then are persuaded that it is their own fault, which it is not.

Flaw #2: No history

The second flaw came to me around 1987. A book on chaos theory, in which I was quoted extensively, became very popular. Journalists called me to ask what the fuss was all about, leading me to write a book, *Chaos, Gaia, Eros,* on the historical context and philosophical significance of chaos theory. Meanwhile, a new course on the history of mathematics was instituted at my university. My colleagues, knowing that was writing a book on the history of chaos theory, offered me this course, and I taught it annually for a decade or more. A friend, Rupert Sheldrake, persuaded me that people learn things better if they are presented in historical order. In the case of mathematics, this is the opposite of the usual, logical order. Armed with this idea, and with my new knowledge of the history of math, I gradually changed all my teaching to a historically based style.

In a historically based program, topics are presented in historical order, so that cognitive prerequisites are available when needed. This avoids the most common obstacles that prevent students from grasping new mathematical concepts.

Flaw#3: No integration.

The historical sequence is crucial, yet not enough. The whole historical sequence should be integrated with the cultural context in which it evolved, providing meaning and motivation for students.

I am convinced that these three flaws are major causes of difficulty that students have in learning math in our school system today. I found fewer failures in my courses after adopting all three remedies—visual representation, historical sequence, and cultural integration—in my courses.

The big test.

The weaknesses in our school math programs today are commonplace, and widely proclaimed. The remedies usually proposed—standardized multiple-choice testing and coaching, word problems, short-question drill and kill, and so on—will do more harm than good. Rather, we advocate graphics, history and integration.

The program of this book

The six chapters of this book are exemplary of the graphical, historical, and integral approach to math. In addition, they illustrate the stages of math as they occurred in the evolution and history of our culture, according to the theory developed by William Irwin Thompson and myself on the five mathematical mentalities: aRithmetic, Geometric, Algebraic, Dynamical, and Xaotic (RGADX).

For example, we advocate teaching geometry before algebra in school. This conforms to historical order, as G precedes A in RGADX. In the history of mathematics, Babylonian geometry evolved into Greek geometric algebra,[12] an essential prerequisite for the Islamic development of rhetorical algebra. Breaking this sequence may be a major cause for math anxiety in our schools.

The chapters of this book are devoted, one each and in order, to the five mentalities, RGADX, and one extra, on Fra Angelico, who wins the prize for the biggest bolt of all.

---

12 (Katz, 1993; p. 64)

1. The Canon of Lespugue

While dynamical systems theory and chaos theory are new branches of mathematics, arithmetic, geometry, and algebra reach back into antiquity. The cognitive style of prehistoric cultures was arithmetic for millennia, whereas the cognitive style of classical civilizations was geometric, an innovation of the ancients, sometimes credited to Thales or Pythagoras. Geometry did indeed flourish magnificently in ancient Greece. But it was within the older Riverine cultural ecology, comprising the cultures of the Nile, Mesopotamia, and the Indus valley, that the ageless aRithmetic Mentality first gave way to the Geometric Mentality. This was a major bifurcation of uncertain provenance; surveying farms for tax purposes is a suspected cause.

As mathematics has evolved from astronomy and music, we may regard archeoastronomy and archeomusicology as the foundation stones of archeomathematics. These two foundation stones were eventually connected in the vision of the *music of the spheres* of the Pythagoreans, circa 400 BCE. [1]

This chapter is an exercise in archeomusicology, one that seeks to demonstrate the existence of the aRithmetic Mentality long before the beginnings of the Riverine cultural ecology. Our goal is to establish the extraordinary development of arithmetic by our Stone Age ancestors.

As the findings that we report here assume a high degree of musical sophistication on the part of our human ancestors before 23,000 BCE, we should begin with a review of some recent archeomusicological discoveries that support this assumption.

---

1   See (Godwin, 1987; Part III) and (Mountford, 1920).

Fragment of a flute, circa 40,000 BCE

At the present time, the discovery of this bone flute provides us with the world's oldest known musical instrument. This artifact is a segment of the femur bone of a cave bear, dated 41,000 to 80,000 BCE. It has two complete holes and two partial holes, one at each end of the broken fragment. It was found in 1997, at a Neanderthal campsite, by paleontologist Ivan Turk of the Slovenian Academy of Sciences, and has been analyzed by Canadian musicologist Bob Fink. Its tuning corresponds to Mi-Fa-Sol-La of the diatonic scale.[2]

A Complete flute, 7,000 BCE

Six complete flutes, and fragments of 30 others, have recently been recovered from burials at Jiahu, in Henan Province, China. The site is firmly dated from 7000 to 5700 BCE.[3] Made from leg bones of the red-crowned crane, the complete flutes have 5, 6, 7, and 8 holes. The best preserved flute has seven nail holes, and has been spectroscopically analyzed. It is apparently tuned to a six-tone or seven-tone Chinese scale. This 9000 year old instrument is still playable, and part of "The Small Cabbage," a Chinese folk song, was played on it and recorded and may be heard at the website for *Nature*.[4]

The presence of these musical instruments indicates that there is a musical culture that goes back to the Ice Age. It is reasonable to assume, therefore, that cave painting, decoration of tools, chanting, and music co-evolved during this period.

In applying the methods of musical arithmetic to measurements of the Venus of Lespugue—a paleolithic sculpture of about 23,000 BCE—we have found that the ratios of lengths of parts of the body to the whole expresses a seven-tone diatonic scale, an ancient scale that is basic to

---

2   For more details, see: http://www.webstersk.cal/greenwic/FLCOMPL.HTM For a skeptical view of this interpretation, see Science News, 153:14 (April 4, 1998) p.21.

3   See (Zhang, 1999)

4   http://www.nature.com

the musical theory of the Vedic Aryans and the ancient Greeks.[5] The lines on the back of the figure may thus indicate a tradition of stringed instrumentation, or some form of Ice Age proto-lute or canon.

The Paleolithic Goddess Figurines

The most surprising aspect of our discovery is the extreme antiquity of the Venus of Lespugue. This is a sculpture carved from mammoth-ivory, 6 inches (147 millimeters) high, from the Gravettian-Upper Perigordian culture, dated about 23,000 BCE, and found at the Des Rideaux site at Lespugue, Haute Garonne, France.[6] The ages of our historical scales have been traced by Ernest McClain back only to early Sumer, around 4,000 BCE.

About 40 similar goddess figures—often called paleolithic Venus figurines, or Aurignacian Venuses—have been found since 1895 in sites of Europe and west Asia, as shown in Fig. 1.5. Theories of their meaning abound in the literature of prehistory since 1967. A very interesting recent summary has been given by Gimbutas.[7] The eight most studied exemplars are those of Lespugue, Kostienki, Dolni Vestonice, Laussel, Willendorf, Gagarino (2 cases), and Grimaldi. Outline drawings of the Venus of Lespugue, enclosed in a rhombus, is shown in Fig. 1.6.

The proportions of rhombi enclosing the figurines provide interesting ratios, like those studied by Villard de Honnecourt for drawings of the human figures.[8] Of the eight figurines the ratios of width to height are:

0.39, 0.33, 0.37, 0.37, 0.44, 0.44, 0.20, 0.28

These numbers seem too closely grouped to be random, and are close to the proportions of the Pythagorean triple triangles: 5-12-13, 12-35-37, and 44-117-125, of the cuneiform tablet Plimpton 322 from

5   See (McClain (1979).

6   See (Graziosi, 1960), (Leroi-Gourhan, 1967), (Marshack, 1972), and (Gimbutas, 1989).

7   See (Gimbutas, 1989). (Leroi-Gourhan, 1967, p. 90), and (Gamble. 1982).

8   See (Bowie, 1959), and (Kayser, 1946).

ancient Babylonia, dated about 1700 BCE.[9] Each of these triples represents the three sides of a right triangle. They are integers, and at the same time, satisfy the Pythagorean theorem. The Plimpton 322 tablet demonstrates an understanding of the Pythagorean theorem of plane geometry, over one thousand years before Pythagoras. Further, the proportions of the Venus figurines indicates a precocious aware-ness of Pythagorean triangle proportions in the Aurignacian period. In short: if these are simply obese human shapes, the obesities are very coincidentally Pythagorean!

Musical arithmetic

The medieval quadrivium—arithmetic, geometry, astronomy, and music theory—is traditionally credited to Pythagoras, and the "Greek miracle" of the 6th century BCE. But the presence of upper paleolithic cave paintings, signs, and musical instruments indicates that this associa-tion of art, mathematics, and astronomy may have its roots in the Ice Age.

By musical arithmetic we mean the music theory part of the quadriv-ium, relating the tones of the scale to the spacing of frets on a stringed instrument, or holes of a flute, and thus to ratios of whole numbers. Pythagoras was reputed to have discovered musical arithmetic from the notes of an anvil struck with a hammer by a blacksmith.

Many different scales have evolved in different cultures and ages. For example, the major and minor modes of modem European music are dif-ferent seven-tone scales. That is, there are seven notes within the span of a single octave. The ratio 2:1 for the octave, and 3:2 for the perfect fifth are widely known. But there are many other correspondences between numerical ratios and harmonic musical intervals. All this is the province of musical arithmetic.

The musical arithmetic of many ancient scales has been recovered by various authors. Pythagoras was known (according to ancient if legend-ary accounts) to have traveled and studied in Egypt, Baby Ionia, and India. The encoding of this ancient musical arithmetic in the dialogues of Plato has been analyzed,[10] and the numerical methods of the Pythago-

---

9   McClain has given a tonal interpretation of Plimpton 322. (McClain, 1978, p. 124).

10 See (Brumbaugh, 1954).

reans graphically reconstructed.[11] The discovery of the musical arithmetic of Pythagoras encoded in the poetry of the Rg Veda pushed back the dating of this sophisticated mathematics by a millennium.[12] We are about to push it back another 20 millennia, but first, we will describe the Hindu-Greek modes.

The Monochord or Canon of Pythagoras

The *monochord* is a one-stringed musical instrument, like the ektar of India. Also known to the Pythagoreans as the *canon*, or otherwise in ancient Greece as the pandoura,[13] the monochord was used as a musical instrument by the ancient Egyptians since 1570 BCE.[14] To the Pythagoreans, it was a scientific rather than a musical instrument; that is, it was constructed primarily for the purpose of experiments in the physics of sound. According to legend, the last words of Pythagoras were an admonition to his disciples to "study the monochord." [15] In these studies were discovered the relations between the harmonic intervals of music theory and the ratios of arithmetic. This discovery is usually attributed to Pythagoras in the 6th century, but is attested in historical evidence only after 300 BCE.[16] Repetition of these experiments on the monochord or canon are described in detail in numerous texts ancient and modern, and are highly recommended as a basis for understanding of the Venus of Lespugue as a canon.[17] We will now describe one scale as a series of whole numbers, and in parallel, as the lengths of the monochord string between frets. The first step in this description is the choice of a reference scale for the lengths of string.[18]

---

11 See (McClain, 1978).

12 See (McClain, 1979).

13 (Levin, 1994, p. 61)

14 (Levin, 1994, p. 71)

15 (Levin, 1994, pp. 71, 96)

16 See David R. Fideler in (Guthrie, 1987, pp. 24-28, 47. 327-328) and (Levin, p. 96).

17 See especially Fideler in (Guthrie, 1987, pp. 24-28), (Levin, 1994, pp. 143-147), and (McClain, 1978, pp. 169-175).

18 For more instruction, see (Terpstra, 1993) and (Fideler, 1993).

A monochord lesson

The relationship between number and tone—common to the Vedic Aryans, Sumerians, Hebrews, and Greeks—involves a reference scale of integers used for defining ratios between the numbers 1 and 2, such as 3/2, 4/3, etc.

The octave, or length ratio 2:1.

Thus we must imagine a monochord with a string length of two meters, say, and a single movable fret. At the top of the string, we inscribe the number "2", indicating two meters, at the midpoint, the number "1", indicated one meter from the bottom, and we put "0" at the bottom, as shown in Figure 1.1. When the fret is located at the "2", and the string is plucked between the movable fret and the base of the instrument at the "0", we hear the lowest note of the instrument, say D below middle C of the piano. And when the fret is moved to the midpoint, at the "1", and the string plucked below the fret, we hear the note D', one octave higher than D, just above middle C. We are going to continue experiments with the monochord, but we will only place the fret between the "2" and "1", and only pluck the string between the movable fret and the "0".

Other reference scales.

As we wish to relate other harmonic intervals—fifths, fourths, thirds, etc.—to ratios of whole numbers, we will find it convenient to replace the "2" by a larger (even) number, and the "1" by half of that number. We consider intervals of integers, [N, 2N][19], as reference scales. The case N=1, or reference scale [1, 2], was described above for the octave interval. By using reference scales of the form [N, 2N], we insure that we always have the number N in the middle of the string, and the number 2N at the top. For example, some reference scales found in the literature use the whole number intervals: [30, 60], [72, 144], [360, 720], and so on.

The integers of the set {N, N+1, N+2, ... , 2N}, inscribed on the fret board, give us reference points between the bottom note, D, and the

---

19 This means the set, {N, N+1, N+2, ... , 2N}.

note one octave above, D'. In general, we inscribe only carefully chosen integers of the reference scale on the body of the monochord, indicating positions for the fret corresponding to notes of the scale to be allowed in musical performance. These favored numbers comprise the reference scale for a given musical scale.[20]

The perfect fifth, or length ration 3:2.

If we mark the monochord evenly with the numbers {0, 1, 2, 3} with "0" at the bottom as before, and '3' at the top in place of the '2", and put the fret at the "2", we hear the note A, a perfect fifth above the bottom D. This is the length ratio 3:2. However, we now have no index for the fret at the octave point, at the center of the fret board. Since we want to be able to sound at will both the octave and the fifth, we will double this range, to {0, 1, 2, 3, 4, 5, 6}. Then we have the fifth at the '4", and the octave at the "3". Our reference scale (indices between the octave and the lowest note at the top of the string) is now [3, 6], as shown in Figure 1.2.

The perfect fourth, or length ratio 4:3.

Let us now choose, as reference scale, the whole number interval [2, 4], which just has the three integers, {2, 3, 4}. We now relabel the monochord fret board, placing the "2" at the midpoint, the "4" at the top, and the "3" half way in between, as shown in Figure 1.3. Sounding the lower segment of the string with the fret at the '3' gives us the note G, a perfect fourth above the lower note D of the open string. We may also find the octave, at the "2". And note that the length ratio of the fourth at the "3" to the octave at the "2" is 3:2, a fifth. That is, a fifth above a fourth is an octave, as 4/3 times 3/2 is 4/2 or 2!

But we no longer have an index for the perfect fifth above the base note. However, if we triple our reference scale to [6, 12], we have harmonized the fifth (at the "8") and the fourth (at the "9"). Note also that

---

20 Usually, these integers are products of the prime numbers 2, 3, and 5. Also, the reference numbers corresponding to tones of a musical scale should be closed under reciprocation. Details are found in many musical texts, but McClain is particularly clear on this theory.

the ratio of lengths from the fourth fret at "9" to the fifth fret at "8" is 9:8, which is the proportion for a typical whole tone of our scale. The ratio 10:9 is also sometimes needed for a slightly smaller whole note, as we will see in the diatonic scale.

The Hindu-Greek Diatonic Scale

Bypassing many complications, we now choose a reference scale adequate for our analysis of the Venus figurine, [72, 144]. This is the smallest reference scale which may be interpreted as a scale of string lengths on a monochord for the diatonic scale. McClain calls this range of integers the *Davidic set*. The heptatonic scale found by the usual Pythagorean method -in rising order (decreasing string lengths, increasing tones)—is thus:

Lengths: 144, 135, 120, 108, 96, 90, 80, 72
Tones: D, e^b, f, G, A, b^b, c, D'.
Ratios: 15/16, 8/9, 9/10, 8/9, 15/16, 8/9, 9/10

This is the *Hindu-Greek Diatonic Scale*, also known as the Dorian Mode of ancient Greece and as Ptolemy's Diatonic Syntonon, and is shown in Fig. 1.4.[21] Here the frequency ratio 15/16 is the reciprocal of the length ratio 144/135, and is the ratio of the frequencies of vibration of the string of the monochord, for the notes D to e-flat, a semitone (half-note) interval. Similarly, 8/9 is the reciprocal of lengths 135/120, and is the frequency ratio for f to e-flat, a tone (whole-note).[22] And so on.

Musical Modes

The pattern of rising intervals—whole tones (T) and semitones (S)—in the Hindu-Greek Diatonic Scale or Greek Dorian Mode—is: STT, T, STT. This notation signifies the rising sequence of small intervals, Semitone-Tone-Tone, Tone, Semitone-Tone-Tone, comprising this scale

---

21 See (McClain, 1976, Chart I, p. xxi, and Chart 3, p. 13), also (Levin, 1994, p. 77).

22 F and f are slightly different notes.

or mode. The first note is taken above as D, but could be any note. Thus our rising scale is approximately: D-D#-F, G, G#-A#-C. This sequence agrees with the Bhairavi Thaat, one of the ten modes of North India.[23]

The Greek Phrygian Mode, also known as the Christian Modus Primus and as the Modern Dorian Mode, has the pattern TST, T, TST.[24] This sequence may be obtained from the preceding by one rotation, that is, starting the scale one note higher. It agrees with one of the Modern Minor Modes, and with the Kafi Thaat of North India.

The Greek Lydian Mode has the pattern ITS, T, ITS.[25] It may be obtained from the preceding by one rotation. It agrees with Modern Major Mode, and with the Bilaval Thaat of North India.

All these Greek modes are usually presented in Pythagorean ratios involving powers of the first three primes only: 2, 3, and 5. Considering the ancient aulos, or Greek flute, as studied from actual surviving examples, the Early Greek Dorian Mode is found in variants involving ratios of powers of the first six primes: 2, 3, 5, 7, 11, and 13.[26]

The patterns, STT, T, STT, etc., are of the form: tetrachord, whole tone, tetrachord. Using only whole tones and halftones, there are only three tetrachords which fit into this pattern and add up to an octave. Therefore, there are nine possible heptatonic (7-tone) scales made according to this pattern. Of the nine, only three are symmetric, that is, have the same tetrachord sequence in both positions. And these three are the Greek Dorian, Phrygian, and Lydian Modes.

According to the rules of Aristoxenos only intervals S and T are allowed in Greek modes, with S exactly twice, while other combinations are found in the (older) scales of North India. Also, S is allowed between the two tetrachords, but there should not be TTTT in the scale, nor in its extension to two octaves.[27] The seven Greek modes satisfying these rules are (in order of rotation):

Mixolydian: STT, S, TTT
Syntonolydian: TTS, T, TTS (symmetric)

---

23 See (Batish, 1989).

24 See (McClain, 1976, p. 61).

25 See (Lauer, 1989).

26 See (Schlesinger, 1939) and (Lauer, 1989, p. 202).

27 See (Mountford, 1920).

Phrygian: TST, T, TST (symmetric)
Dorian: STT, T, STT (symmetric)
Lydian: TTT, S, TTS
Ionian: TTS, T, TST
Aeolian: TST, T, STT

There are no other modes following these rules.

Measurements of the Venus of Lespugue

We oriented the outline drawing of the Venus, front view, as shown in Fig. 1.7. Choosing horizontal lines (frets) across the figure in the most obvious places (using a computer drawing program) and measuring the distances on the Davidic reference 1.8 scale [72, 144] we have introduced above and in the second column of numbers in Figure 1.4, we obtain the results of Fig. 1.8. The accuracy of our measurements is within one half of a reference unit, which is less than a millimeter in the scale of the actual figurine, 147 millimeters. The correspondence with the Hindu-Greek (Greek Dorian) Diatonic Scale is within 1 unit, as shown in Fig. 1.8, except for the seventh tone. This result we call the *Canon of Lespugue*.[28]

In Fig. 1.8, two extra notes of the chromatic scale in just intonation, b (at fret 86) and a^b (at fret 102), have been interpolated among the Dorian frets for comparison. Comparing the two scales, Venus at the left and Dorian at the right in Figure 1.8, we may note two main differences.

First, note that the subdominant perfect fourth, G, and the dominant perfect fifth, A, occur in both scales. The subdominant fret of the Venus scale coincides with the top of the cusp between the breasts. The dominant fret of the Venus scale coincides with an incised line at the top of the pubic triangle. In between, in the Venus scale, we have noted a line defined by the bottom of the breasts, corresponding to a-flat, the tritone. This most discordant note does not occur in the Dorian scale. We may interpret this line as a fret for occasional ornamentation, and not part of the Canon of Lespugue: it is an extra fret.

---

28 Compare with Fig. 1.1, held upside down. Regarding the note symbols, see (McClain, 1976, pp. 33, 125, 127).

Secondly, the seventh note of the Canon of Lespugue at coordinate 84 is audibly flat, relative to the seventh note of the Greek Dorian Mode, c, at 80. It is also slightly sharp relative to the b of the chromatic scale in just intonation, so we might denote this b+ or c-. Of the seven notes of the diatonic scale, this is the only one which differs significantly from the Greek Dorian Mode. An interval greater than a whole tone is suggested, in contradiction to the (later) rules of ancient Greece.

We may compare three scales (the Dorian, the Lespugue, and the aulos measured by Schlesinger) in Davidic length coordinates:

Dorian: 144, 135, 120, 108, 96, 90, 80, 72
Lespugue: 144, 134, 119, 108, 96, 91, 84, 72
Aulos: 144, 131, 118, 105, 92, 86, 79, 72

Note that the 7th note of the Lespugue, at 84, is closer to the 6th note of the aulos. Otherwise, the Lespugue is closer to the theoretical Dorian than is the actual aulos of ancient Greece!

Further support for our musical interpretation of the Goddess of Lespugue is provided by the incisions on the back of the figurine, which look very much like the strings of a lute, gathered to a single peg or nut, with the buttocks serving as the resonating chamber, as shown in Fig. 1.9.

Conclusion

We may wonder how a sculpture, about six inches high, could be measured so accurately, its main divisions placed accurately within a millimeter! The musical arithmetic might have been worked out experimentally on a much larger scale. The monochords of the Pythagoreans, about two meters long, come to mind. Some mechanism of reduction in scale, such as a nomograph painted on a cliff side or large flat rock face, may have been employed.

Historians and prehistorians have greatly underestimated the intellectual capacities and cultural traditions of pre-agricultural societies. The bias of the myth of progress that is basic to an industrial civilization, and the bias of behaviorism that was basic to twentieth century social sciences, seem to have blinded scholars to the significance of the artifacts they unearthed. The age of highly developed artifacts can be taken back to 30,000 BCE at least.

We are therefore faced with the need to upgrade our estimation of the cultural advancement of paleolithic people. To the great arts of the paleolithic period, including the fabulous paintings of Lascaux and the Goddess figurines, we must now add an advanced musical arithmetic previously ascribed to Pythagoras, the Vedic Aryans, or the Old Babylonians. This supports the suggestion of McClain that advanced arithmetic preceded writing, and demands a rethinking of prehistoric mathematics and music. It would seem useful to proceed with measurements of the other figurines and artifacts of paleolithic times, in the style of Alexander Marshack and Edmund Carpenter, and early myths in the style of Hertha von Dechend and Gorgio de Santillana, in search of further clues to the quadrivial pursuits of prehistory.

Movable Fret

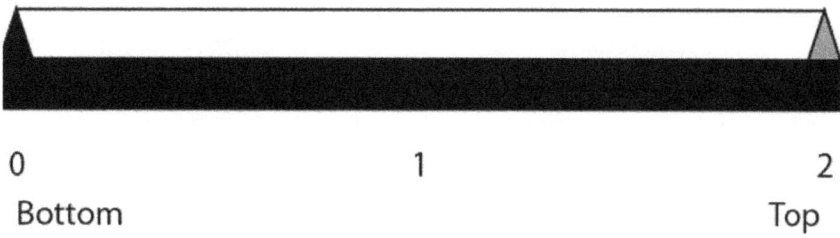

| 0 | 1 | 2 |

Bottom   Top

FIGURE l.Ia. The monochord set to sound, D, the base note.
The string (upper black line) should be plucked anywhere to the
left of the movable fret (grey) that is able to slide along a track
in the heavy base (black). The reference scale in this setup is [1 ,
2].

Movable Fret

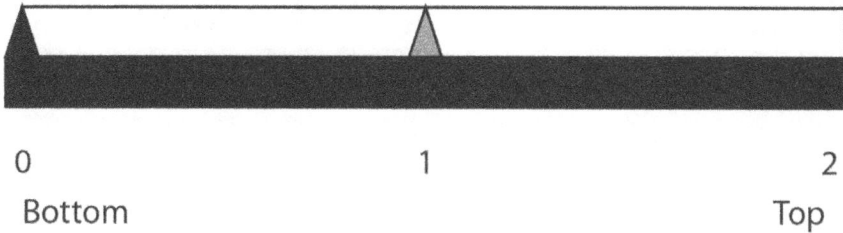

| 0 | 1 | 2 |

Bottom   Top

FIGURE l.lb. The monochord set to sound the octave, D'.
above the base note, D. The string should again be plucked to the
left of the movable fret.

Movable Fret

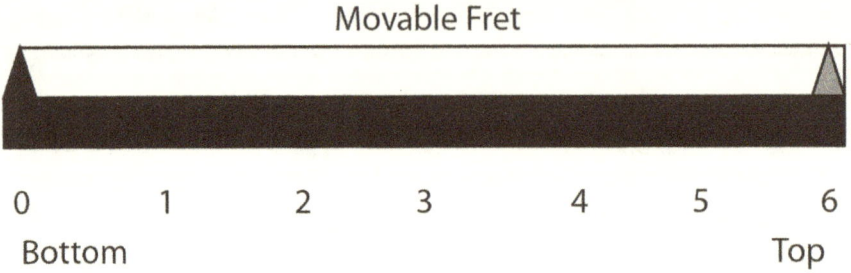

| 0 | 1 | 2 | 3 | 4 | 5 | 6 |

Bottom   Top

FIGURE l.2a. The monochord set to sound, D, the base note. This is identical to Fig. l.la, except that the reference scale in this setup is [3, 6].

Movable Fret

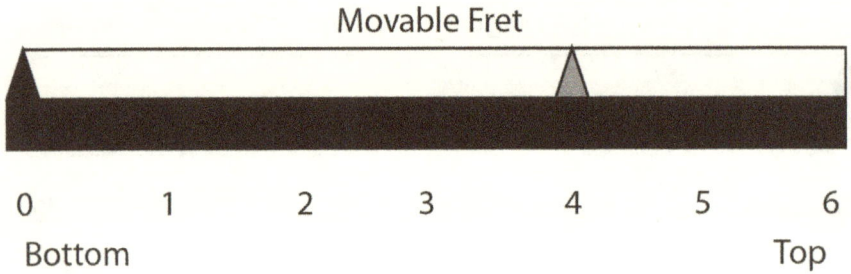

| 0 | 1 | 2 | 3 | 4 | 5 | 6 |

Bottom   Top

FIGURE l.2b. The monochord set to sound, A, a perfect fifth above the base note. D. The reference scale is again [3, 6].

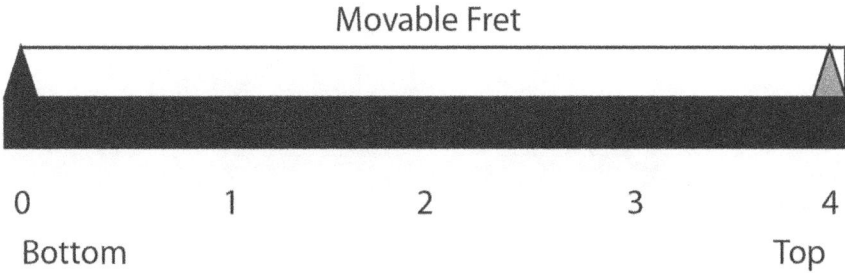

FIGURE 1.3a. The monochord set to again sound, D, the base note. The reference scale is now [2, 4].

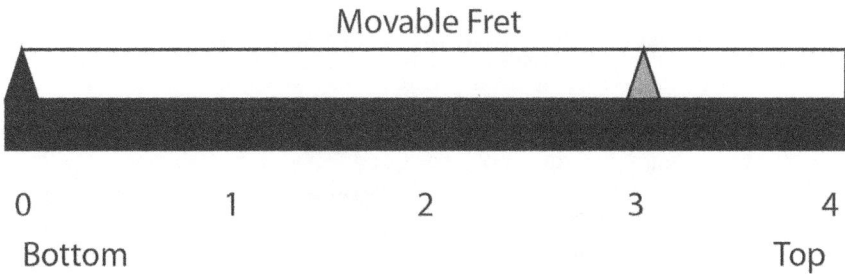

FIGURE 1.3b. The monochord set to sound, F#, a perfect fourth above the base note, D. The reference scale again is [2, 4].

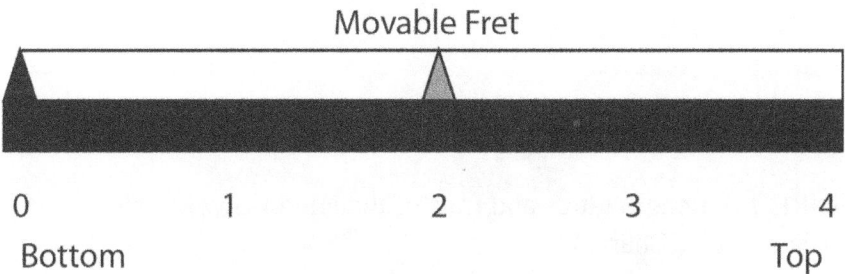

FIGURE 1.3c. The monochord set to sound D', an octave above the base note, D. Note that the length ratio of F# to D' is 3:2, a perfect fifth.

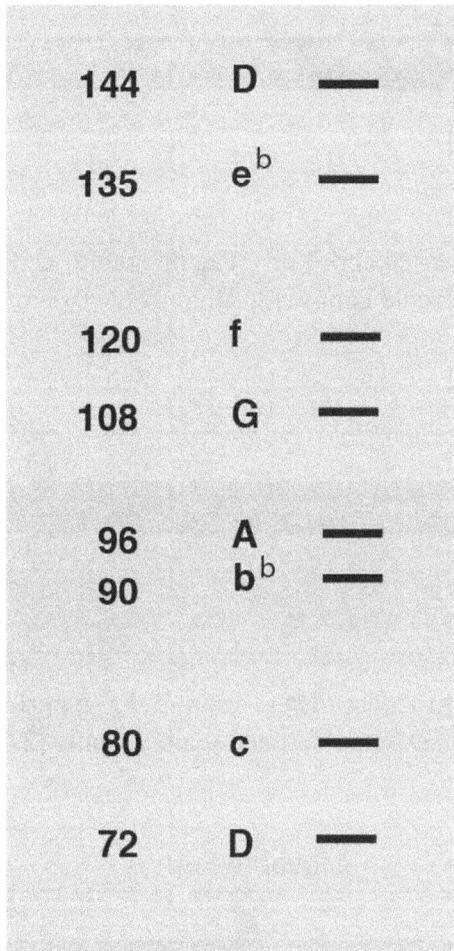

FIGURE 1.4. Length ratios and frets of the Hindu Greek scale (McClain. 1976, Chart I)

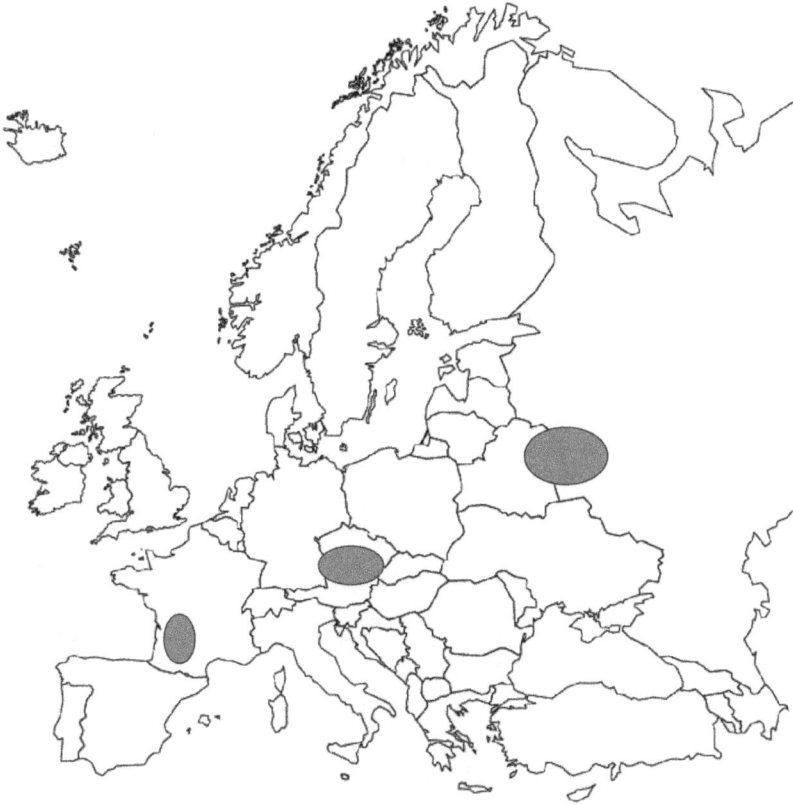

FIGURE 1.5. Map of Europe showing the distribution of paleolithic goddess figurines. (Gamble, 1982)

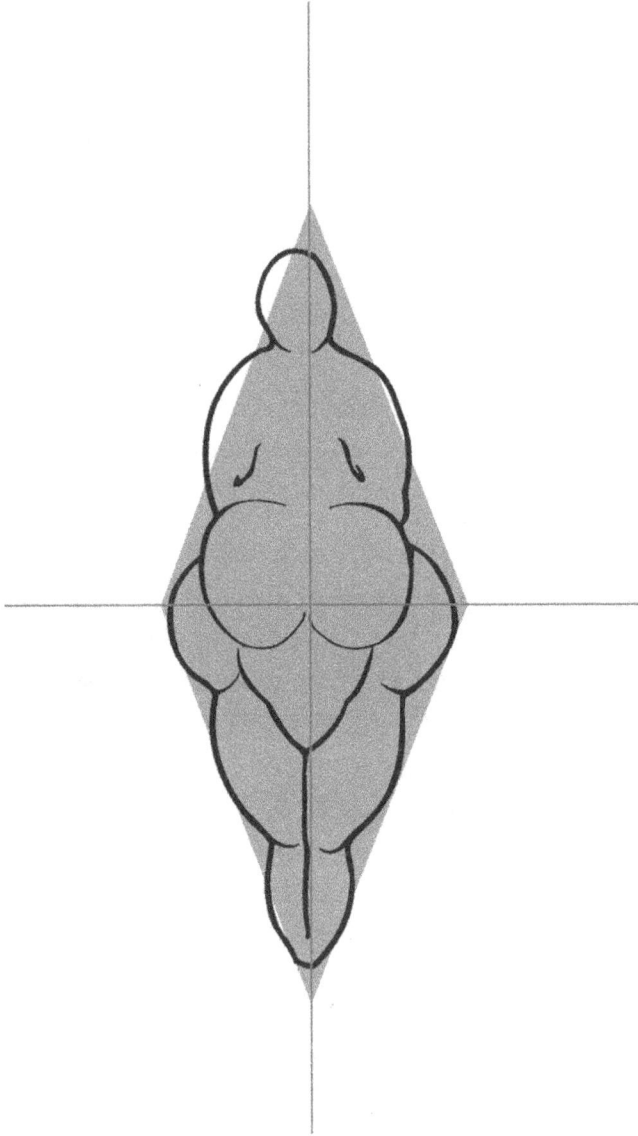

FIGURE 1.6. Venus of Lespugues, enclosed in a rhombus. (Leroi Gourhan, 1967, p. 92)

FIGURE 1.7. Outline drawing of the Venus of Lespugue, front view. Sketched from a photo the actual sculpture. (Gimbutas, 1918; p. 163.)

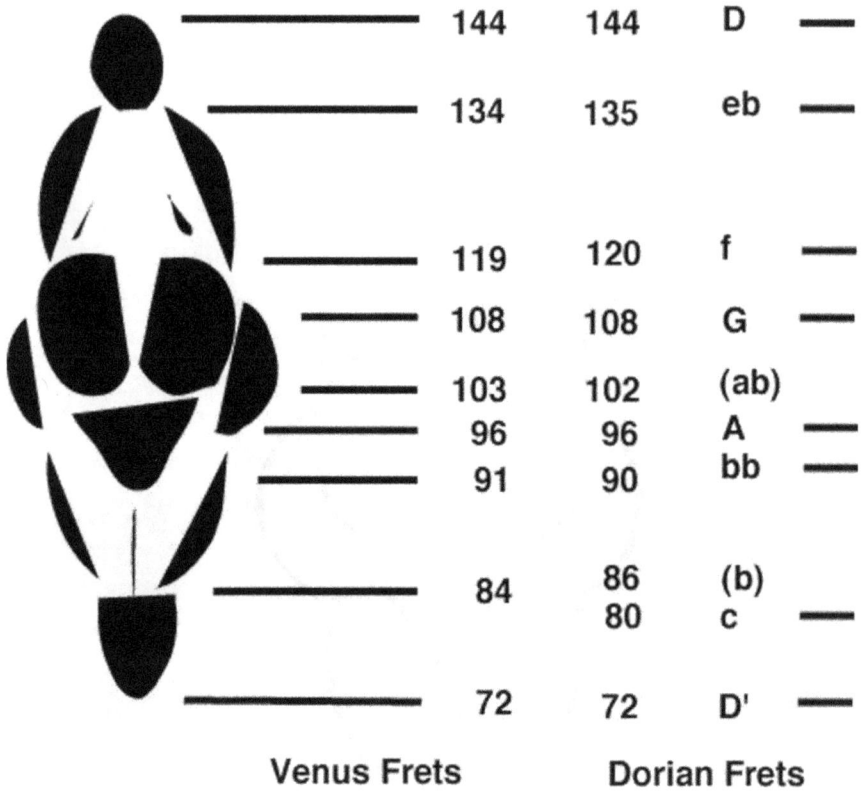

FIGURE 1.8. The Canon of Lespugue alongside that of the Greek Dorian Mode.

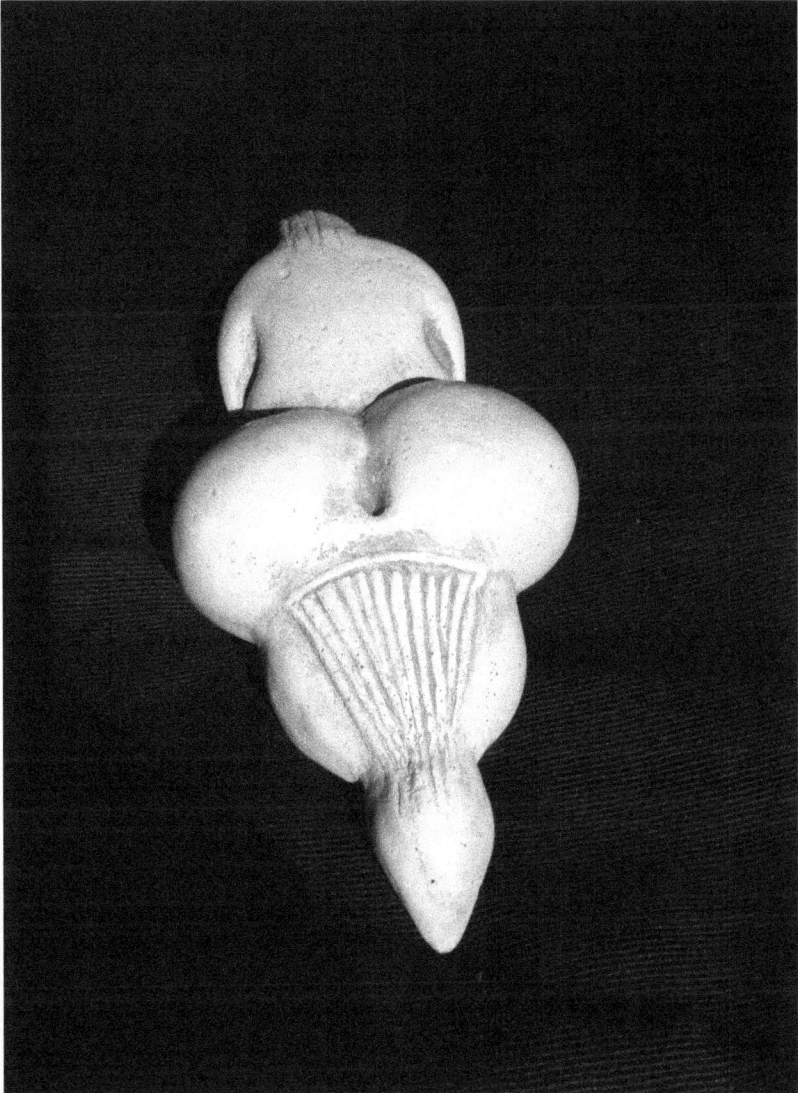

FIGURE 1.8. Back of the Venus of Lespugues. Photo of a reproduction.

2. THE BETHELS OF SCOTLAND

The aRithmetic Mentality stretches over an immense span of time. Beast, birds, and fish have arithmetic skills. Newborn humans are able soon after birth to add and subtract small numbers.[1] Hominid arithmetic evolved into the aRithmetic Mentality certainly 100,000 years ago, perhaps earlier. By 23,000 BCE, as we have seen in the first chapter, our paleolithic ancestors were quite advanced. In comparison, the Geometric Mentality is a new arrival. In this chapter, we will see that this way of thinking is also perhaps older than we might have thought.

The R/G shift from the aRithmetic into the Geometric Mentality took place around 500 BCE, according to the usual chronology. Beginning with the innovative world travellers, Thales and Pythagoras, a creative wave was set in motion which culminated in the Academy of Plato, around 380 BCE, and was recorded by Euclid, around 300 BCE. Essentials of Babylonian and Egyptian mathematics were rearranged by Euclid into a formal system, the backbone of our mathematics too this day.

The *Elements* of Euclid has been the second most published book of all time, after the Bible, and the most important math text in all history. It builds to a climax in the final Book XIII, in which the five Platonic solids are constructed. These constructions may be understood as the epitome of the Geometric Mentality.

And yet, 2500 years before Euclid, these five Platonic solids were already well-known and appreciated by the megalithic people of the British Isles—a bolt from the blue—who left us evidence of their knowledge in the form of small carved stones, called the bethels of Scotland. This chapter we tell the story, and follow the thread of these ideal geometric figures—known variously as cosmic figures, regular solids, Platonic solids, and so on—from prehistoric Scotland to the Baroque Austria of

---

1   (Dehaene, 1999).

Kepler. At the end, we will return to this unanswerable question: What were the megalithic people doing when they plucked the cosmic figures out of the blue?

The Megalithic People

Megalithic (i.e., large stone) monuments are scattered about Europe. The large circles of standing stones—Callanish in Scotland, Avebury and Stonehenge in the south of England—are perhaps the best known. They are the remnants of a scientifically advanced prehistoric culture that we call the megalithic people.

These people migrated across southern Europe, one branch to Scandinavia on a northwesterly track, another via Malta to Brittany and Britain on a southwesterly track. Megalithic monuments are evidence of the high culture of the these people. They set up megaliths as they migrated from place to place, and standing stones still mark their prehistoric way. The megalithic circles of Britain date from 3000 to 1500 BCE.

Callanish

About half of all the stone circles of Britain are in the northern part of Scotland. On the island of Lewis in the Outer Hebrides there is a collection of a dozen or so sites named Callanish 1, II, III and so on. Callanish I is the second largest stone circle in Britain after Avebury. These are the largest megalithic remains in the entire world. Callanish 1. functions as a solar and lunar observatory, testifying to the mathematical and astronomical sophistication of these people, who inhabited a topography and latitude ideal for celestial observations.[2] In the first century BCE, the historian Diodorus Siculus wrote about Hyperborea, a mythic place in the north:

> The moon as viewed from this Island appears to be but
> a little distance from the earth and to have on it promi-
> nences like those of the earth which are visible to the eye.
> They say that the god visits the Island every 19 years...

---

2   See (Schultz, 1983/1990), and (Thom, 1977)

> There is also on the Island both a magnificent sacred
> precinct of Apollo and a notable temple...[3]

Certainly this description fits Callanish better than Stonehenge or any other megalithic monument, and the visits of the god probably refers to the 18-year cycle of eclipses, the Saros cycle discovered by the Babylonians.[4]

Lunar Archeoastronomy

Callanish was a very sophisticated lunar observatory, as discovered by Alexander Thom.[5] Observations in prehistoric observatories were made along the horizon. In fact, Sir Norman Lockyer said that the horizon was the telescope of prehistoric people. Given a good horizon, as for example on the roof of a tall building or pyramid, or on a treeless and flat island, one easily observes the exact point on the horizon at which a star rises or sets.

In the case of the sun, the daily rising point on the eastern horizon (in the northern hemisphere) moves north in the fall, stands still at the winter solstice (this word means sun-still), and moves south again in the spring, reversing again at the summer solstice. Observing the sunrise thus provides a rough calendar of the seasons.

In the case of the moon, the daily rising point in the east may be noted relative to the rising point of the sun. The lunar rising follows the annual cycle of the sun, but also changes relative to the sun. There are several lunar cycles, interrelated in a complicated fashion, and these are helpful in predicting eclipses. Evidently the Callanish archeoastronomers understood them.

The Carved Stone Balls

Returning now to the Platonic solids of Euclid's *Elements*, one should

---

3   (Hawkins, 1973, p. 57)

4   The Saros period of 18 years, 10 days, and 8 hours, is useful in predicting both solar and lunar eclipses. There are about 29 lunar eclipses and 41 solar eclipses in a Saros period. Saros is Chaldean for repetition. See (Rey. 1952; p. 138) and (Schultz, 1983/1990; pp. 30, 99-111.)

5   (Thom, 1971)

realize that we do not find these in the cultures of Babylonia nor Egypt. However, it seems they were known to the megalithic people of Scotland by 3,000 BCE, from whom hundreds of carved stone balls, *bethels*, have come down to us.[6] They are carved out of the hardest available rocks, and seem to have been handled a great deal. Most were found in the northeast of Scotland, at about the same latitude as Callanish.

Some 411 bethels located in 36 museums (and various private collections) around the world have been catalogued at the National Museum of Antiquities of Scotland. Of these, most are about the same size, with diameters of about 70 mm, like a grapefruit. Dorothy Marshall classified them as Type 1 up to Type 9. Type 1 has three knobs; Type 2 has four knobs, and so on. About half of them look very much like Platonic solids.[7]

What could these bethels have been used for? According to one theory, they were carved to receive a thong which bound them to a club, turning them into weapons. Another suggestion is that they were part of a ball game, and still another, they were part of a system of standard weights. REcently it has been proposed that the bethels were used as ball bearings for moving the standing stones of Stonehenge.[8]

In any case, these stones remained in common use up until about 1000 BCE. Each stone most likely had a very long lifetime, because they are made of diorite and basalt—extremely hard stones that are difficult to work, but last forever. The labor necessary to work the stone and maintain their perfect geometry suggests that they served some important purpose. They have been found in connection with neolithic and epipaleolithic tools, axes and hammers in quarries, dating from a time when people knew well how to work stone, do sculptures in hard stones, and raise huge monuments. That these stones were created within a culture capable of constructing astronomical markers for the sky and horizon suggests to us that they also served as some form of cosmological expression.

Mathematical Litherature

There are many speculations about the purpose of the carved balls, as

6   Bethel, in Hebrew, means "house of God."

7   See (Marshall, 1976) and (Marshall, l983).

8   http://www.kpbs.org/news/2010/nov/12/nova-secrets-stonehenge/

noted above. But these ignore the meaning of the very elaborate carvings. We speculate that they were the result of mathematical discoveries, and therefore part of the litherature[9] of mathematics. Some of them have many more knobs than any Platonic solid, but all five cosmic figures are extensively represented among the bethels found near the Callanish observatories.

The Five Cosmic Figures

Here are the data for the cosmic figures. The order and data for the first four are taken directly from Plato, as described below. The fifth is from Euclid, the definitive authority.

- tetrahedron, 4 faces (equilateral triangles), 4 vertices (at which 3 faces meet),
  - octahedron, 8 faces (equilateral triangles). 6 vertices (4 faces meet)
  - icosahedron, 20 faces (equilateral triangles), 12 vertices (5 faces meet)
  - cube, 6 faces (squares), 8 vertices (3 faces meet)
  - dodecahedron, 12 faces (pentagons), 24 vertices (3 faces meet)

Next we give cardboard-and-tape constructions of each, our own expression of the constructions known to the Pythagoreans. For the first three, we:

- construct an equilateral triangle on paper,
- make copies on cardboard (4 for the tetrahedron, 8 for the octahedron, or 20 for the icosahedron),
- tape them together.

For the cube, we:

- construct a square on paper,
- make 6 copies on cardboard,
- tape them together.

---

9   Litherature: our neologism for preliterate knowledge carved in stone.

For the dodecahedron, we:

- construct a pentagon on paper,
- make 12 copies on cardboard,
- tape them together.

Of these constructions, only the pentagon is difficult, and that was a major triumph of the Pythagoreans, around 400 BCE. The official straight-edge-and-compass constructions of the equilateral triangle, square, and pentagon are described in detail, and proved, in Euclid's *Elements*, Books I and IV.[10]

From Prehistoric Scotland to Kepler

It is commonly believed that the Megalithic people were instructed by wandering priests from ancient Babylonia or Egypt. Perhaps this is true. We certainly might be able to trace connections between the solunar observatory at Callanish and the astronomical knowledge of Babylonia. But in which direction was the vector of diffusion? The Babylonians had very accurate observations on the relations between the different phases of the moon, for example, the Saros period discussed above. These accurate observations are made by recording the point on the horizon at which the sun and moon rise and set.

Callanish provided a perfect observatory, particularly due to its latitude, for making these observations, as the wandering of moonrise along the horizon is exaggerated near the poles.

We may begin our story, then, in Ancient Babylonia, leaving open for the time being the question of the role of the carved stones of Scotland, and the origin point of the cosmic figures.

Ancient Babylonia

History may begin in Sumer,[11] as the distinguished Sumerologist

---

10 Computer-graphic illustrated text and animations may be found on ou website at www.visual-euclid.org.

11 Pronounced "Shoomer". See Thorkild Jacobsen in (Cotterell, 1980; p. 72).

Samuel Noah Kramer claimed in his important text of 1959, as there the earliest writing system evolved.[12] But mathematics began deep in prehistory. While the arithmetic, astronomy, and calendrics of the Sumerians and their successors, the Babylonians, were highly evolved, their geometry was not. Surviving texts indicate a rudimentary development of plane geometry—for example, the area of plane rectilinear figures, or the Pythagorean theorem. Simple geometric algebra was known, including a geometric construction for solving simple quadratic equations, such as: $x^2 + 2x = 4$. Solid geometry was less developed, but methods for calculating the volume of a solid figure were known in some cases, e. g., a truncated square pyramid. There is no surviving text on the cosmic figures.[13]

Ancient Egypt

The origin of the math and sciences is credited, in early Egyptian myth, to the god Thoth, known as Hermes by the Greeks. He taught humans most aspects of civilization: cosmology, writing, astronomy, astrology, arithmetic, geometry, music, chemistry, alchemy, theurgy, and magic.[14] Practical arithmetic and plane geometry were highly developed, and applied to economics, astronomy, surveying, architecture, temple orientation, and so on. But as far as historical evidence is concerned, we know little of their solid geometry. This situation is very similar to that of the Ancient Babylonians. Although the Great Pyramid provides us with evidence of great skill in practical solid geometry, we have no record of the five cosmic figures in Ancient Egypt. The Egyptian and Greek tradition on the origin of geometry was first set down by Herodotus, then repeated by other writers, including Heron, Diodorus Siculus, and Strabo.

The Egyptian tradition that geometry was the gift of the god Thoth was repeated by Socrates in Plato's *Phaedras*. The Greek word *harpedonaptae* (rope stretchers) was coined for the geometers of Egypt. Rope stretching in Egypt is attested as early as 2300 BCE.

Herodotus tells us that Sesotris (Rameses II, circa 1300 BCE) distributed land among all the Egyptians in equal rectangular plots, on which

---

12 (Schmandt-Besserat, 1996)

13 (Katz, 1993; Secs. 1.5, 1.8, 1.9)

14 (Fowden, 1986)

he levied an annual tax; when therefore the river swept away a portion of a plot and the owner applied for a corresponding reduction in the tax, surveyors had to be sent down to certify what the reduction in area had been. 'This, in my opinion, he continues, 'was the origin of geometry, which then passed into Greece.[15]

In sum, we may regard ancient Egypt and Scotland as the early homes for archeogeometry. Meanwhile, Babylonia and Scotland were home to very early archeoastronomy and associated numerics.

The bethels of Scotland suggest to us the possibility that Scotland preceded both Babylonia and Egypt in the development of early mathematics.

Prehistoric Italy

In the 1870s, 26 dodecahedral objects of Celtic origin were found in Italy. And in 1885, near Padua, there was discovered an actual regular dodecahedron of Etruscan origin, dated 1000-500 BCE.[16] These artifacts are most likely the detritus of migrating peoples and knowledge from remote origins in Scotland, Babylon, or Egypt to Greece.

Thales, Pythagoras, and the Pythagoreans

Thales (624-547 BCE), the first of the presocratic philosphers, was among the first of the Greeks to write on mathematics, science, and philosophy. It is said that he travelled extensively, studied in Egypt, and then introduced geometry to Greece. With Thales, geometry became a deductive system. He was able to measure the height of a pyramid, and gave proofs of at least five theorems now recorded in Book I of Euclid's *Elements*. But there is no record of his interest in the cosmic figures.

Pythagoras (572-497 BCE) was the next important Greek geometer after Thales. Pythagoras travelled extensively, including a long stay in Egypt studying mathematics and astronomy with the priests. On returning to Greece, he furthered the program of Thales to develop geometry and arithmetic into a formal deductive system, and founded a religio-mystical community in the south of Italy.

---

15 (Heath, 1960; p. 121)

16 (Heath, 1960; p. 160)

Proclus says, in his Commentary on Book I of Euclid's *Elements*, that after Thales, Pythagoras transformed the study of geometry into a liberal education, examined the principles of the science from the beginning, and probed the theorems in an immaterial and intellectual manner.

Pythagoras even now is regarded as a most important figure in the history of mathematics. He is said to have discovered that the rational numbers (ratios of whole numbers) are not enough. For example, the diagonal of the unit square (according to the so-called Pythagorean theorem, known long before Pythagoras) is the square root of two. The Pythagoreans (followers of Pythagoras) discovered (to their dismay) that this number is not any ratio; it is irrational.They also discovered the construction of the cosmic figures.[17]

When Proclus writes "construction," he means the ruler-and-compass construction of the equilateral triangle, square, and pentagon, and the cardboard~and-tape construction of the five cosmic figures, as described above.

The treatment of the cosmic figures in Euclid is their first appearance since the bethels of Scotland, with the exception of the prehistoric Etruscan objects noted above. There may be other, unknown exceptions, comprising a continuous trail of stone from Scotland to Italy, without which, the achievements of Pythagoras, Plato, and Euclid would have been impossible. Thus, our trail seems to lead from megalithic prehistory, to ancient Babylon and Egypt, to Thales and Pythagoras, and finally to Plato and Euclid.

Plato and the Platonic Academy

Plato (429-347 BCE) became a follower of Socrates. During a visit to Sicily, he met the Pythagorean, Archytas of Tarentum, and became a Pythagorean himself. Returning to Athens, he founded the Academy around 385 BCE, over the entrance to which was written (according to tradition), "Let no one ignorant of geometry enter here." His writings, a number of dialogues regarded as the foundation of all Western philosophy, have been divided into three phases by historians. The *Republic*, written around 400 BCE in the middle phase, describes Plato's mathematical curriculum for the Academy: arithmetic, plane geometry,

---

17 (Heath, 1960; p, 141)

solid geometry, astronomy, and music theory. Plato complains that solid geometry is not well enough developed.[18] We will be concerned here with the *Timaeus*, the dialogue which divides the middle and last phases. Here Plato introduces the cosmic figures, henceforth to be also known as the Platonic solids. He describes not only the figures, but also their crucial significance in the structure of the universe: a theory which prevailed from 400 BCE up to 1600 CE.

The *Timaeus* is a discussion of four persons: Socrates, Timaeus, Critias, and Hermocrates. It begins with a review by Socrates of a discussion on the preceding day. This concerned the constitution of the ideal State and its citizens. Then Critias tells the famous story of Atlantis, which was told to his great-grandfather by Solon, one of the seven sages. [21] (Numbers in brackets are page numbers of the Stevens translation.) Then Timaeus is asked to begin the feast with a description of the creation of the Universe. [28]

He tells how God, because he was good, made the world after an eternal pattern. He brought order into the world, and soul and intelligence. [30] The world is composed of fire and earth. [31] Being solids, these two elements require two more, water and air, to bind them. [32] The world is a sphere [34] with the soul in the center. [35] The proportions used in the creation of the world were those of the Pythagorean Tetractys and the Diatonic scale. [36] Within the sphere are the circles of the seven planets. Within this frame God made the body of the universe. [37] Then he made time, a moving image of eternity. [38] And then he made the four species of animals, [39] gods, birds, sea, and land animals. [40]

The gods made man and the lower animals, and God made the human souls of the same four elements as the body of the universe, along with part of the soul of the universe. [41] Then he sets in motion the process of incarnation, and reincarnation, of these human souls into mortal bodies. [42] The created gods make these mortal bodies of the four elements. [43] As a person becomes a rational creature through education, his human soul moves in a circle in the head (a sphere) of his mortal body. [44] The head obtains sight from a reaction between the light of the eyes and the light of day. [45] From sight we derive number and philosophy. [47]

Now we come to the nature of the elements. [48] In the creation process there are three natures: an intelligible pattern, a created copy,

---

18 (Katz, 1993; p. 49)

and the space in which creation proceeds. [49] Space can receive any form, that is, the impress of any idea. [50] The elements are affections of space, produced by the impression of ideas. [51] The four elements took shape in space, and God perfected them by form and number.They are solid bodies, and all solids are made up of plane surfaces, [53] and plane surfaces in turn are made of scalene and isosceles triangles. Three elements are made from equilateral triangles, the fourth from isosceles triangles. [54] The first and simplest solid is the tetrahedron, the second is the octahedron, the third is the icosahedron, and the fourth is the cube. God used a fifth solid to delineate the universe. [55] The elements are shaped as follows: earth as the cube, water the icosahedron, air the octahedron, fire the tetrahedron. [56]

At page [56], four of the five cosmic figures have been described, including all the details we have given above. Plato obviously knows all about these four solids, and knows also that there are five, but the dodecahedron is not described in detail in the Timaeus.

After page [56] there are details on how space can change from one element to another, and on the number of solids which make up a visible piece of matter. This comprises a kind of atomic theory. The Timaeus ends on page [92]. We will move on now to Aristotle's theory of matter.

Aristotle

Plato described matter as a substance consisting of a mixture of elements, each of which consists of atoms, which in turn are simply space pressed into the form of a cosmic figure. To Aristotle, matter was a combination of substance and attributes. This theory is called hylomorphism, after hylé (Greek for "forest".) The substance is actual stuff made of atoms, and the shape of the atoms determines the attributes. This theory,] is fundamental to the Catholic ritual of the Eucharist.

Euclid

Euclid systematized all the arithmetic, geometry, music theory, and astronomy of his time. His works became the basis of the quadrivium, half of the school curriculum that was standardized in the Christian middle ages. The other half was the trivium: grammar, rhetoric, and dialectics. Together, the quadrivium and the trivium comprised the seven classical arts.

Euclid's Elements

The *Elements* comprise, in thirteen books, the arithmetic, geometry, and geometric algebra of Plato's Academy. Until very recently, it was used as a textbook in our schools, and every famous mathematician of note studied from it. It was replaced by modern texts at about the same time that math anxiety and the math avoidance reaction became wide-spread.[19] As described in our Introduction, the failure of the math curriculum in today's schools is a serious disorder of our society, and here we are pointing to one of the crucial issues. But let us return to Euclid's text.

The first six of the 13 books of Euclid's *Elements* are devoted to plane (that is, two-dimensional) geometry. These six books comprise 148 propositions. Of these, 48 are constructions (ending in QEF in the Latin editions of Euclid, or *that which was to have been done*), the rest are theorems (ending in QED in the Latin, or *that which was to have been proven*). That is, for Euclid, a construction is to be done, while a theorem is to be proven. In fact, the main motivation of the theorems, in the beginning, was to prove that the construction actually work. That is, if we follow the steps correctly to construct a square, then in the end, we have a figure that is actually square.[20]

In other words, the core of the *Elements*, and Euclid's goal, is a set of constructions. When Euclid gives a construction, let us say to bisect an angle, he then gives the constructive steps through which the angle is bisected, and then afterwards proves that what you've done is actually what you wanted to do. In our perspective, most of the work—the formal system of axioms, definitions, theorems. proofs, and so on—is intended to establish that the constructions do what you want them to do. The constructions themselves represent the essential heritage of Babylonian

---

19 English edition of Euclid's *Elements*.

20 See Heath's elegant Preface to the first edition of his classic.

and Egyptian geometry, while the formal system represents the Greek contribution: the innovation of Thales, and its working out within the Greek cultural matrix.

The cusp between these two traditions is actually the R/G shift (or bifurcation, as we now like to say) from the aRithmetic to the Geometric Mentality. If we had to place our finger on the crucial moment of bifurcation, this might be the Pythagorean discovery that the square root of two is an irrational number. For here, geometry transcends arithmetic. What shifts at this cusp is the entire mentality of the cultural ecology. The entire mindset and cognitive style of the Ancient Greek world moves from an algebraic to a geometric style, as manifest in literature, music, philosophy, politics, and so on.[21]

After the first six books of Euclid on plane geometry come four books of arithmetic. Then, Books XI, XII and XIII return to geometry with the three-dimensional case, and Book XIII culminates with the construction of the five Platonic (or cosmic) solids.

The Five Cosmic Figures in Euclid

The first three figures were known by the ancient Egyptians, and the dodecahedron by the Pythagoreans. The icosahedron is usuallgf attributed to Theatetus, around 380 BC, at Plato's Academy.[22] All these were very important to Euclid. It could be argued that they were the goal of the Platonic Academy, and the reason why it said over the door, "Let no one ignorant of geometry enter here." Later, schools posted a notice, "Let no one come to our school, who has not first learned the *Elements* of Euclid."

The five regular solids were thought to be the building blocks of the actual stuff of the universe, that even empty space itself is a construction—like Tinker Toys or Legos—of these Platonic blocks. This was taken seriously all the way up to the time of Kepler, who proposed that the universe was designed by God from these five Platonic solids, as described below.

Theatetus, around 380 BCE, wrote the first systematic treatment on

---

21 See (Thompson, 2004).

22 (Heath, 1908)

the five cosmic solids.[23] He was probably the first to understand that there were only five regular solids.[24] The results of Theatetus, and his follower Aristaeus, around 320 BCE, were improved and completed in Book XIII of Euclid's *Elements*, which is relatively independent of the earlier Books. Each of the five cosmic figures is inscribed within a comprehending sphere, and its edge value—that is, the ratio between the length of an edge and the diameter of the comprehending sphere—is established.[25]

Greeks after Euclid

Soon after Euclid, his results on the edge values of the cosmic figures were extended by Apollonius (250-175 BCE) and still later by Hypsicles (2nd century CE).[26]

Kepler

When Plato, in the *Timaeus*, assigned the first four cosmic ligures to the elements, he also said that God used the fifth, the dodecahedron—known to the megalithic people of Scotland, the prehistoric Celts and Etruscans, and the Pythagoreans, looks like a soccer ball—in the delineation of the universe.[27] This hint was expanded by Kepler (1571-1630), in his earliest work, the *Mysterium Cosmographicum*, of 1596. Toward the end of his career, in 1621, Kepler revised this work. The revised edition includes an illustration showing his model of the universe, in which a sphere comprehends a cube, which comprehends a sphere, which comprehends a tetrahedron, in its sphere a dodecahedron, in its sphere an icosahedron, which comprehends a sphere.[28] This is shown in Figure 2.2. A sequence of edge values from Euclid's Book XIII determine the diameters of six concentric spherical shells, which carry the orbits of the planets as they circle the sun.[29]

---

23 (Heath, 1908/1956; v.3, p. 438)

24 (Katz, 1993; p. 87)

25 (Mueller, 1981; Ch. 7)

26 (Heath, 1956; p. 439)

27 (Jowett, 1892; v.2, p. 35)

28 Many details may be seen at our website, www.visual-kepler.org.

29 (Katz, 1993; pp. 373-379)

Conclusion

It is evident from the astronomical alignments of the stone circle at Callanish that this megalithic culture had a model of the solar system which is essentially spherical. We are suggesting that before the carved stone balls were carved in very hard stone there were some earlier experimental and impermanent carved clay balls (not surviving) that were the result of efforts to model the three-dimensional geometry whereby the celestial sphere revolves around the Earth, a geocentric model, and the wanderers, especially the sun and the moon, had wavy paths that could be traced on these mud balls. Thus the stone balls are artifacts of mathematical astronomy, just as are the Platonic solids of Euclid and Kepler.

The existence of the cosmic solids carved in hard stone, together with the tradition of the five Platonic solids from Euclid to Kepler, seem to indicate that there is some common ancestor that predates the cultural migrations of the megalithic people, and cultural diffusion of the megalithic knowledge. It does not seem that there was a reverence for the cosmic solids in ancient Egypt, nor in Babylon. They survive only in stone, in the far North of Scotland some 2,000 years before Pythagoras, and again in Italy, a few centuries before Pythagoras.

Since the bethels date from the time of a general Indo-European diaspora, with the Aryans moving toward India and the Proto-Greeks down into the Balkans, one can reasonably conjecture that there also was a third branch, the ancient Hyperboreans discussed by Diodorus Siculus,[30] that went north and west.

---

30 (Diodorus Siculus, 1935, pp. 37-39).

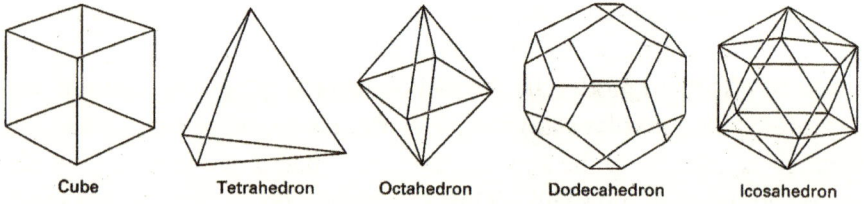

Cube   Tetrahedron   Octahedron   Dodecahedron   Icosahedron

FIGURE 2.1. The five cosmic figures (Katz, 1993, p. 87)

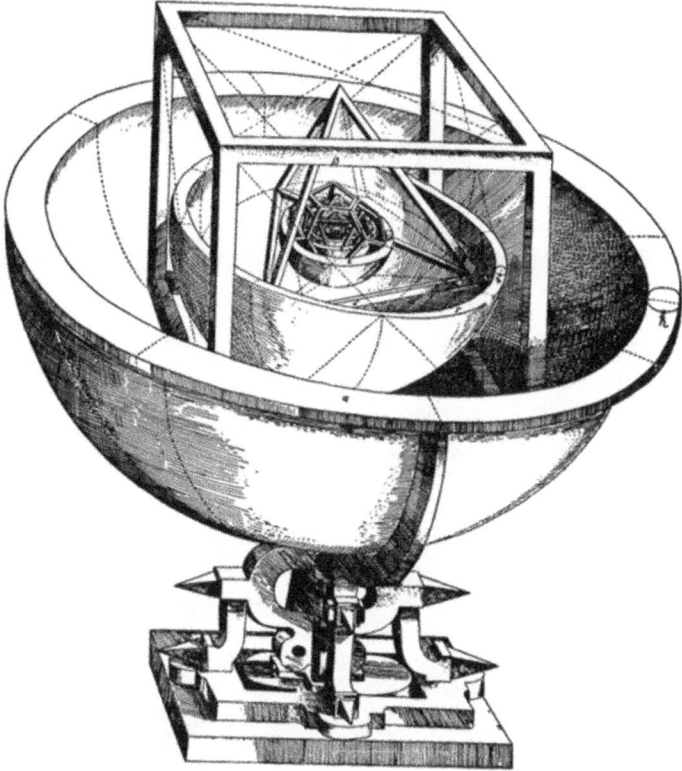

FIGURE 2.2. Kepler's model. The frontispiece from Kepler's Mysterium Cosmographicum. (Wikipedia Commons)

3. MATHEMATICAL CALLIGRAPHY

The G/A shift, in which algebra as we know it was born, was a complex dynamical event.

Introduction

The Algebraic Mentality was dominant for eight hundred years. Here we analyze its birth as the result of a cultural/chemical diffusion and reaction.

Understanding the historical sequence, R/G/A, may be the key to understanding, and thus curing, the worldwide epidemic of math anxiety, so we have subjected it to a full analysis in the style of complex dynamical systems thinking.

The reaction-diffusion paradigm

In the reaction-diffusion paradigm, chemicals diffuse through a physical substrate, and also react with each other, creating a spatial pattern. The chemicals are called morphogens (as they create forms) or reactants. Reaction-diffusion has been used to model some of the mysterious phenomena of biological morphogenesis, such as:

- How does the leopard get its spots?
- How does an egg turn into a chicken?

In this chapter we will use chemical reaction-diffusion as a model for cultural reaction and diffusion. So, metaphorically speaking, we will view the origin of algebra as the result of a reaction between different reactants diffusing through world cultural history, a reaction product that then diffuses through the Mediterranean cultural ecology.

The three-layer model

In our reaction-diffusion model for cultural morphogenesis we will focus on three layers. In order of chronological development, they are:
- the cosmic system, or spirituality,
- the number system, or numeracy, and
- the writing system, or literacy.

These three layers are similar to those of the stratigraphic interpretation of the sacred texts, especially those of Sanskrit, Hebrew, and Arabic literature.[1] In these exegetical traditions, a base layer of literal interpretation supports an intermediate layer of mythic interpretation, on top of which rests a stratospheric layer available only to the spiritual initiate: a sort of triple palimpsest.

Regarding the relative antiquity of these three systems, the cosmic system is first evident in the cave art and pictographs of the old stone age, lets say from 50,000 BP (before present).[2] The number system—as we have seen in our first chapter—is manifest in the Upper Paleolithic, or about 35,000 BP, and writing systems about 6,000 BP.

The cultural histomap

We wish to view these three knowledge systems—spiritual wisdom, mathematics, and literature—as overlays upon a geographic map. Adding motion for change with respect to historical time, this model provides an animated map for the human collective consciousness in which all knowledge resides and evolves. We refer to this kind of representation as a histomap.

So we are now thinking of a cultural histomap having three groups of dimensions: spatial dimensions for geographical places, a temporal

---

1   For Arabic, see (Schimmel, 1984, p. 90).

2   (Lewis-Williams, 2002; p. 39)

dimension for historical time, and conceptual dimensions holding three layers of ideas.

Following the reaction-diffusion paradigm, cultural/conceptual/mental chemicals diffuse in this histomap, and natural evolution occurs gradually, or occasionally, catastrophically. New cultural forms emerge, as in the reaction-diffusion models used in mathematical biology to account for the formation of a leopard's spots in utero. But we are interested instead in the formation of spirituality, numeracy, and literacy, emerging as distinct layers in the conceptual space—the very foundations of human consciousness as emergent phenomena in a complex dynamical system.

We will begin with a review of each of these three reactants, one at a time, in chronological order: spiritual signs, number signs, and letters. Our story of the evolution of algebra begins with the spiritual systems.

Spiritual systems, sacred calligraphy

It is well known that prehistoric pictographs and cave art all over the world carried supernatural associations: sacred, spiritual, cosmological, and so on. We may call such symbol schemes spiritual systems, and the embodied art of their production and contemplation, sacred calligraphy.

As pictographs evolved into prewriting and writing systems in Sumer, Egypt, China, and Mesoamerica, some of these spiritual associations survived. Without doubt the apex of the art of sacred writing was achieved in the hieroglyphics of ancient Egypt. And as phonemics evolved and simplified into syllabaries, alphabets, and new number systems, the new symbols inherited these supernatural associations, along with the literary and numerical layers of signification. These were later developed into gematria and Kabbalah.

As the diffusion of these cultural morphogens radiated outward from Egypt, it infused the early alphabets. While there is no evidence of sacred calligraphy in ancient Greece, there are arguments for a spiritual layer in some Presocratic texts,[3] and in the late antique Neoplatonic and Hermetic corpi.[4] However, the earliest historical record we have of spiritual

---

3   (Bernal, 1987)

4   (Fowden, 1993)

practices based on the letters of the alphabet is relatively recent: the Kabbalistic practice involving the twenty-two letters of the Hebrew alphabet and the ten number symbols of the Indian number system. This literature begins with the *Sefer Yetsirah*, or *Book of Creation*, written somewhere in the interval 200 to 600 CE.[5] And we speculate that from this Hebrew source of biblical exegesis—perhaps transmitted to Western European monastic centers from the Desert Fathers—evolved the spiritual practice of Latin and English calligraphy among medieval Christian monks, and, by 700 CE or so, the spiritual practice of Arabic calligraphy within early Islam.

As Judaic, Christian, and Islamic traditions encountered one another in Iberia, a new syncretic impulse helped to bring forth the culture and literature of the Kabbalists of medieval Europe, one in which a new and highly developed angelology—similar to that of Persian traditions—became articulated in the *Zohar*. Similar syncretic movements of cultural evolution occurred in China, Japan, and elsewhere, and continue even to this day.

At the heart of this meditation upon a sacred text is the idea that each letter or numeral signifies a spiritual state, and that meditation on the esoteric meaning of a sign while drawing it produces that exalted state. For this reason, Muslims resisted the Western mechanical printing of books: "They hold that their scriptures, that is, their sacred books, would no longer be scriptures if they were printed."[6] The slow, embodied, calligraphic production of a word thus induces a sequence of spiritual states, like Tai Chi or Sufi dance. Certain words, then, have been found by experience to ascend Jacob's angelic ladder to heaven. Indeed, this idea is explicit in the Jewish and Islamic mystical literature.[7]

The conventional theory of scholars whose mentality is not within this contemplative spiritual ethos is that calligraphy is simply a craft of writing in a monumental, decorative, or artistic style, evolved from metalworkers in Mesopotamia and Egypt. Perhaps it is useful, in order to recognize both exoteric and esoteric modes of cultural transmission,

---

5   (Scholem, 1941, p. 75)

6   From a letter of Ogier Ghiselin de Busbecq, written in 1560. (Lewis, 2002; p. 118).

7   (Schimmel, 1984)

to contrast exoteric and esoteric forms of calligraphy. The Old Syriac Gospels of the 5th century CE are regarded as outstanding examples of exoteric calligraphy.[8] In either case, exoteric or esoteric, the evolution of the well-attested art of sacred calligraphy plays an important role in our story of the origins of algebra, as we shall see.

Number systems

Contrary to popular belief, it now appears that numbers are older than writing. Tally sticks have been found dating from deep prehistory, and it seems to be firmly established that the first writing systems evolved out of the number systems of counting tokens that were used in the context of inventories, accounting, urban administration, bookkeeping, calendrical timekeeping, and the like.[9] After writing emerged, it absorbed the number system. The combined system of writing and numbers later separated again into two layers, and we return to this story below.

Writing systems

The largest bifurcations of world cultural history, as described in *Chaos, Gaia, Eros*,[10] are: agriculture, the wheel, and the chaos revolution. In the second of these, the wheel, writing and patriarchy emerged together about 6,000 years ago. The developmental sequence of this one bifurcation event—the step-by-step emergence of its various aspects and the cause-and-effect relations among them—is unknown. One possible sequence has the pottery wheel first, then its adaptation as the cart wheel, which empowered the urban revolution—the chariot wheel giving advantage to militant patriarchy and support for the priestly elite—and then writing systems. Another sequence, championed by Leonard Shlain, has writing before patriarchy.[11] David Diringer placed writing above all in importance, and wrote that it represented an immense stride forward in

---

8   (Healey, 1990, p. 50)

9   See (Schmandt-Besserat, l996; p. 7) and (Senner, 1989; pp. 8-9; Ch. 2).

10 (Abraham, 1994)

11 (Shlain, 1998)

the history of mankind, more profound in its own way than the discovery of fire or the wheel.[12]

So an evolutionary process begins to take shape: from agriculture to literature, as it were. First the farms and goods, tokens and numbers to keep track, then the wheel, carts, farmers' markets in towns, writing systems, more sophisticated arithmetic, geometry and geometric algebra to tax the farms, and finally algebra.

Our story of the evolution of algebra continues here with the creation of writing systems. Our interest now is to extract enough data from the history of writing to detail the diffusion of writing on a histomap.

From pictographs to the alphabet

The prehistorical record of pictographic pre-writing systems begins with Upper Paleolithic art, circa 45,000 BCE. Leroi-Gourhan has written extensively on the evolution of the sign system—primitive realistic, quadrangular, brace-shaped, claviform, tectiform, vulvar, and late realistic forms—which survived into the mesolithic art of Europe and Africa.[13]

The subsequent work of Gimbutas traces the further evolution of these sign systems into the pre-writing systems of Neolithic Europe, "the sacred script of the Goddess."[14] The earliest scripts of the Indus valley are still undeciphered, but some scholars, such as Steve Farmer, have argued that the images on the seals are not truly writing but a system of narrative icons[15]—more analogous to the Tarot than to a phonetic system. Pictographic systems are also found in pre-urban Mesopotamia, as well as in predynastic Egypt.

In the 1990s, Denise Schmandt-Besserat argued that the cuneiform system of writing evolved from tokens, and that the markings on these tokens had their roots in neolithic culture.

> The immediate precursor of cuneiform writing was a system of tokens. These small clay objects of many

---

12 See (Diringer, 1962, p. 19) and (Donald, 1991).

13 (Leroi-Gourhan, 1967, pp. 201, 513-5l6)

14 (Gimbutas. l989)

15 (Farmer, 2002)

shapes—cones, spheres, disks, cylinders, etc.—served as counters in the prehistoric Near East and can be traced to the Neolithic period, starting around 8000 BC. They evolved to meet the needs of the economy, at first keeping track of the products of farming, then expanding in the urban age to keep track of goods manufactured in workshops. The development of tokens was tied to the rise of social structures, emerging with rank leadership and coming to a climax with state formation.[16]

Eventually, under the influence of transcultural trade and commerce, writing systems came into wider use, and eventually these traders—whether the West Semites in caravans on land or the sea-trading Phoenicians—consolidated the transformation of the system from syllabic to alphabetic. It was this system that was then taken over by the Greeks, who put it to good use in creating the literature of poetry and philosophy upon which the civilizations of the Mediterranean cultural-ecology were based.

The better number system

After the establishment of writing, number signs were absorbed into the writing system. For example, within the cuneiform writing system of Mesopotamia, a relatively advanced number symbol system was embedded. And after the emergence of the alphabet, letters were used as number signs as well as phonemic signs. What will become two layers is, at this point, just one.

But shortly before the Common Era, a rather different system emerged in India, which evolved with impressive rapidity into the modern system. According to Diringer, the Indian numerals evolved from the Aramaic alphabet, which reached India in the 7th century BCE. At this point, the two layers separated. And with the Indian numerals, we see a cone-like pattern of cultural diffusion from India, this time with greater speed of diffusion. We have a cyclical chemical reaction here: from numbers, to writing, to alphabets, to better numbers. And now, finally, to algebra.

---

16 (Schmandt-Besserat, 1996)

The birth of algebra

Muhammed ibn-Musa Al-Khwarizmi is generally acknowledged as the father of algebra. In the early 9th century CE, he was the leader of the Bayt al-Hikhma (House of Wisdom, an official institution of the Caliphate of Baghdad, and was a key figure in the intellectual development of early Islam.[17] His most important books:

• *On the Hindu Numbers*, which survives only in a Latin translation, and
• *An Abridged Treatise on the Jabr and Muqabala Calculation* (Al-kitab al-mukhtasar fi Hisab al-jabr wa-l-muqabala)

The latter was one of the first algebra texts in Arabic. Our word algebra comes from al-Jabr in this title. As if by coincidence, these two works reflect the influx diffusion of the number system from the Hindus into Islam, and the outflux diffusion of the new algebra from Islam to the West.

In our reaction-diffusion metaphor for the birth of algebra, the Hindu number system diffuses, as a morphogen, into the tradition of sacred calligraphy of Islam. While this was the home ground of ancient Babylonian arithmetic and geometric algebra, the ancient knowledge had been lost. But at the same time, around 800 CE, Euclid's *Elements* arrived from Byzantium.

Euclid's *Elements*, the most important text of mathematics of all time, was created in Alexandria around 300 BCE, went on the road around 600 CE after the destruction of the Alexandrian Library, rested in Byzantium, and came to Baghdad around 800 CE.

The effort of translating Euclid 's refined version of geometric algebra[18] from Greek and Syriac to Arabic stimulated the mathematicians of the Bayt al-Hikma to make use of their newly liberated alphabet for a fresh expression of the old methods.

17 (Sayin, 1989)

18 As described in the previous chapter, geometric algebra was a laborious method of solving equations in ancient Babylon.

Thus three cultural-chemical morphogens diffused into Baghdad at about the same time: sacred calligraphy, the Hindu numerals, and Euclid 's *Elements*. A cultural-chemical reaction resulted in the precipitation of algebra, which then diffused outward. The Hindu numbers dissolved the attachment of the symbols of the alphabet from numerics, thus liberating them for a new level of significance: the unknown variables and operations of geometric algebra.

Three vectors converged on Baghdad in his time. Further, it is certain that al-Khwarizmi knew the recent translations of Euclid's geometric algebra into Arabic, to which he refers in his *Abridged Treatise*. He may also have seen a translation of Diophantus, who had continued the Babylonian arithmetic tradition in Greek antiquity.

To this we may add the radical surmise that al-Khwarizmi was aware of a vestigial survival of the original Babylonian geometric algebra, which no doubt had been the basis for the geometric algebra of Euclid, and the symbolic arithmetic of Diophantus.[19]

Thus a commodious vicus of recirculation, from Babylon to Alexandria and back to Baghdad, with a catalytic agent from India, gave birth to algebra.

In summary

Putting all this together into a chronological sequence on our histo-map:

• Ancient Babylonia had a tradition of geometrical algebra, that is, algebraic problems solved geometrically, by 2000 BC.
• Thales and Pythagoras learned this tradition during their travels, stimulating Greek geometry, around 600 BC.
• Greek geometry was organized by Euclid as a justification and proof of the constructions of Babylonian geometrical algebra, around 300 BC.
• Euclid's *Elements* reached Baghdad and was translated into Arabic around 800 CE.
• At about the same time, the Indian number system arrived from

---

19 See (Neugebauer, 1962, p. 146) and (Klein, 1968, p. 127)

India, and sacred calligraphy arrived from the Christians. The numerical layer was displaced by the spiritual in a signilicant cultural phase transformation.

• In this new cultural phase, al-Khwarizmi's efforts to understand Euclid, bolstered perhaps by a residuum in situ of the old Babylonian tradition, gave rise to algebra.

This story is more of a reaction-diffusion event than a bolt from the blue into the mind of an isolated genius. And yet, as far as we know, the actual chemistry took place entirely within the mind of al-Khwarizmi.

The rise of algebraic symbolism

The algebra of al-Khwarizmi was not yet the algebra we know. It was still rhetorical, that is, the variables and operations were denoted by letters and words, rather than by the signs and symbols which characterize algebra today. It may be interesting to note the step-by-step evolution of the modern symbols. The plus and minus signs appeared in Germany around 1500 CE, that is, about three centuries after the arrival in Europe of the Hindu numbers. A few decades later, the equal sign appeared in England. The symbolic notations of Diophantus reemerged in a work of Viete in 1591.[20] And in 1637, it all came together in the *Discours* of Descartes, with modern notations for variables, exponents, operations, and so on.[21] This process may be regarded as an enlargement of the number system, and further reduction of the burden placed on the writing system by mathematics. With the diffusion of the several cultural morphogens into the fresh soil of Europe, a fresh burst of chemical reactions brought forth the Renaissance. The ancient and medieval developments of geometry, arithmetic/algebra, astronomy/astrology, medicine/alchemy, Kabbalah/magic were rebom.

---

20 (Klein, 1968, p. I5I; Katz, 1993, p. 339)

21 (Cajori, 1928; Katz, 1993, p. 399)

Conclusion

There is little doubt about the concurrent arrival in the new and spirited cultural milieu of Baghdad, early in the 9th century, of:

- sacred calligraphy
- the Hindu numerals
- Euclid's *Elements*.

The connection between this concurrence and the birth of algebra in the same time and place as a cultural/chemical reaction is plausible perhaps, yet not conclusive. One objection immediately comes to mind: Why did we not have this explosion in India, where the liberation of the Devanagari syllabary (that is, the Sanskrit alphabet) from its numerical burden came a century or two earlier?

One possible reason is that Hindu polytheism still embraced concrete idols, but Islam's great clearing away of idolatry prepared for the elimination of the concrete in favor of the invisible and abstract. Also, the entrancement with sacred scripture was much stronger in the medieval Jewish, Christian, and Islamic cultures than in India, where the tradition was older.

Our understanding of this cultural shift involves the idea of the numerical layer as a sort of electrochemical insulator or barrier between the phonemic and the spiritual layers. With the separation of the numerical from the phonemic layer, the spiritual and abstract mathematical states came into catalytic interaction with the alphabet, and rhetorical algebra arrived as a bolt of chemistry in the blue.

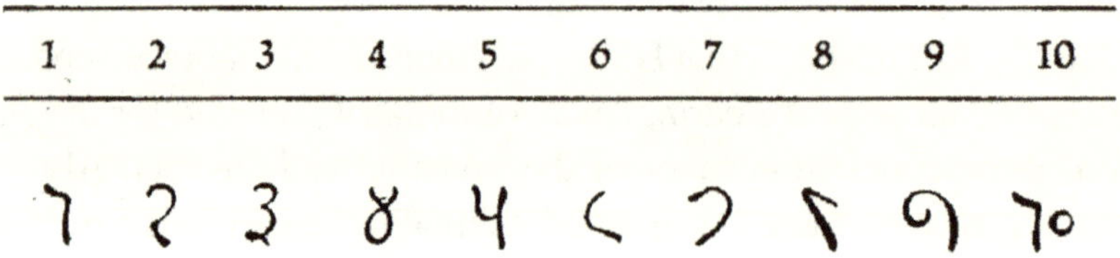

1	2	3	4	5	6	7	8	9	10

Figure 3.1. Gwalior Numerals, ca 800 CE. From Joseph, p. 82.

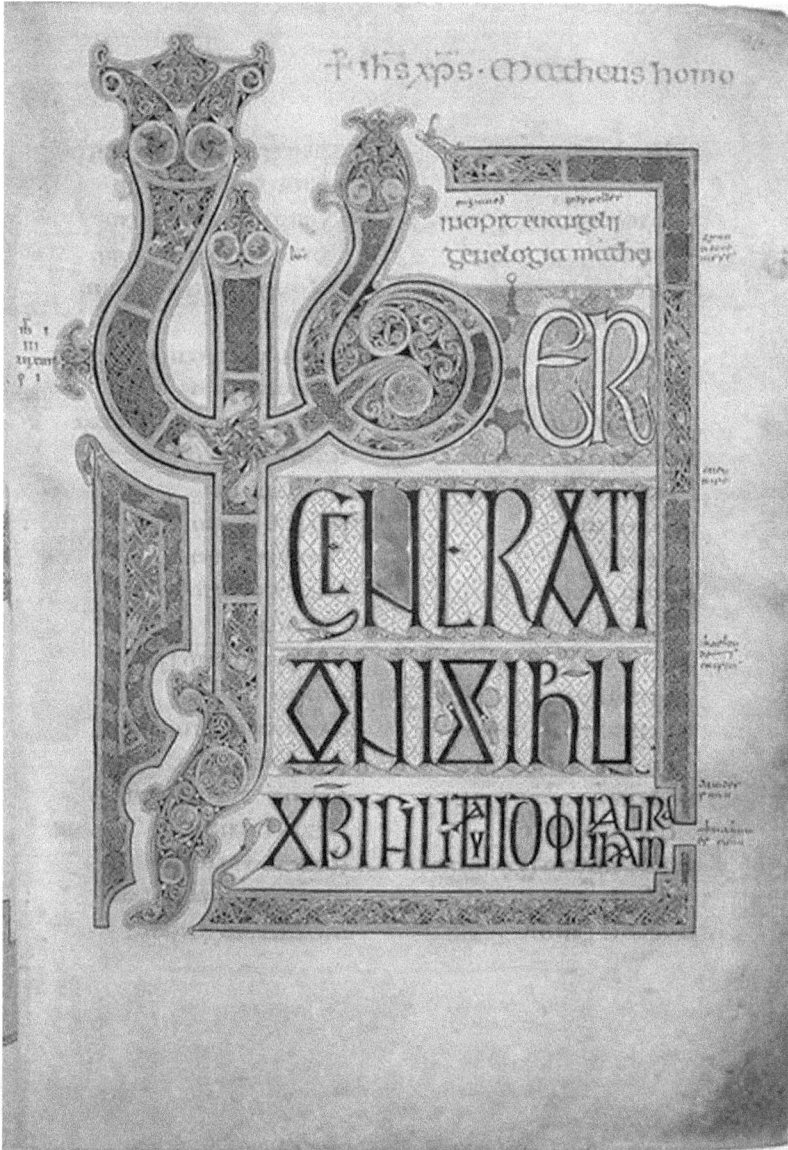

FIGURE 3.2. Latin/English calligraphy. From the Lindisfarne Gospels, Folio 27r. Wikipedia Commons.

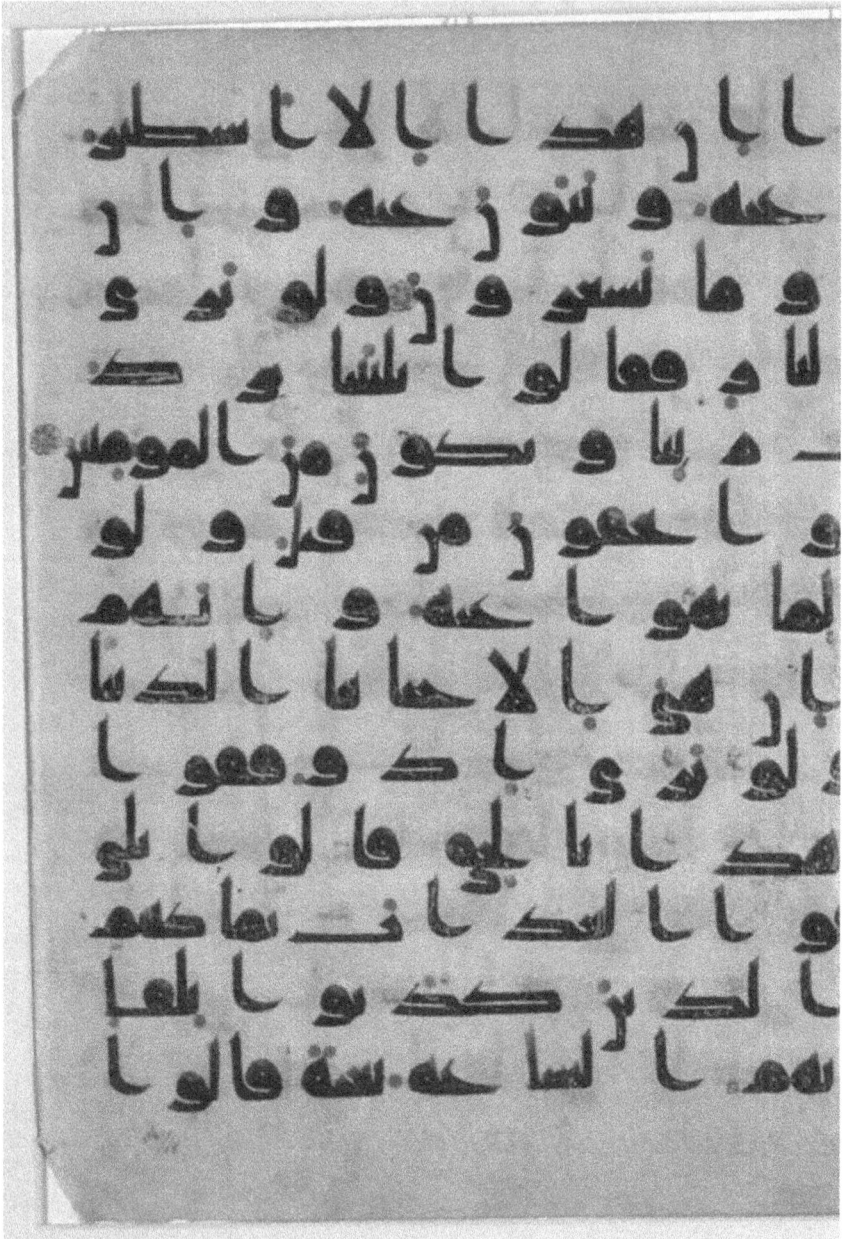

FIGURE 3.3. Arabic calligraphy. From the Koran. Wikipedia Commons.

4. THE GEOMETRY
OF ANGELS

The next major shift in our primary sequence, the birth of the Dynamical Mentality around 1600, is the subject of the next chapter. Here we will tell the story of a radical bolt from the blue, in which Fra Angelico anticipated in 1438 the early stages of the complex topologies of the 20th century! This is an archetypal bolt, in that the Renaissance priest saw in his meditations, and then painted on canvas, the three-sphere, a mathematical object that would not emerge into the mathematical literature for 500 years.

Dante and Giotto

We begin with some background on the history of art: Dante and Giotto. In the European tradition, Dante (1265-1321) is regarded as one of the consummate poets. He has been regarded as marking the end of the Middle Ages, and was followed soon by Petrarch (1304-1374), who was among those who triggered the Renaissance, and by Boccaccio (1313-1375), the early Humanist. All were influenced by the troubadours of Provence.[1]

Dante was composing the *Divine Comedy* from 1307 until his death. In the Angel Sphere of Canto XXVIII of the Paradiso, according to the very convincing argument of Mark Peterson,[2] Dante describes the three-sphere in words. His vision is itself a notable bolt from the blue, a precognition of 20th century mathematics. Dante was geometrically sophisticated for his time.[3] Giotto (1277-1337) was a contemporary and

---

1   See (Smith, 1973, p. 18).

2   (Peterson, 1979)

3   (Crosby, 1997, pp. 171-173)

friend of Dante. He painted Dante's portrait, and was an early advocate of natural perspective in paintings.

Fra Angelico

Fra Angelico (1400-1455) was inspired by Giotto, and participated in the revival of the geometric art which characterized the Renaissance. His was "one of the most innovative and responsive pictorial minds" of his time.[4]

Following Mark Peterson's insight that there is a visionary geometry in Dante,[5] we are going to maintain that the three-sphere is represented in paintings of Fra Angelico around 1435. His representation utilizes the construction of the three-sphere from two families of nested tori, a construction known to modern mathematicians as the *Hopf fibration* after its discovery by the topologist Heinz Hopf around 1930. This construction is shown in Figure 4.5.

Nimbus or Halo?

An ellipse is a circle seen obliquely, and an early application of the ancient Greek theory of conics to painting is the elliptical representation of hat brims and halos by Giotto, around 1310. The halo was seen as a golden disk attached to the heads of saints. But in Fra Angelico, the dinner-plate halo is replaced by the spherical nimbus. No matter the angle of view, the nimbus is invariably shown as a golden disk bounded by a black circle: round, not oval. As Fra Angelico was mathematically sophisticated (one of his paintings is based upon an Alberti grid), we have to assume that his nimbus represents a semi-transparent ball of golden light. Imagine then a spherical plastic bubble, which is transparent from the outside, but semi-opaque when seen from the inside.

---

4   (Spike, 1997, p. 11)

5   Dante's vision may have roots in the Sufi cosmology of Ibn Arabi. See (Palacios, 1968). See also Ibn Arabi's "Diagram of Mystical Cosmos" in (Turner, 1995, p. 40), and compare it to the Rosa Candida of Dante's Paradiso, Canto XXXI1.

Fra Angelico's first spherical nimbus appeared on the Linaiuolo Madonna of 1433.[6]

One may contrast the elliptical halos, suggestive of golden dinner plates, in Fra Filippo Lippi's Barbadori Altarpiece of 1437, with the nimbi of Fra Angelico of about the same time.[7] While the circle represents perfection in a two-dimensional world, the sphere comprehends the entire three-dimensional cosmos.

The angels

The angels of Fra Angelico, perhaps inspired by visionary experiences seen during meditations and prayers, seem to fly upward through dimensions. They create three-dimensional space by beating their wings: two dimensions plus time make three dimensions. This implied motion is an alternative to perspective as a symbolic form representing three-dimensional space.[8] Perfect circles are divine, according to Plato, and so also are spheres. We should understand this geometrical perfection as a suggestion of the perfection of sanctity, expressed in the circular halo and spherical nimbus.

The three-sphere

We will explain how to construct the one-sphere, then the two-sphere, and then the three-sphere. The one-sphere is simply a circle. The two-sphere is the ordinary two-dimensional sphere. But the three-sphere is a counterintuitive three-dimensional object. The three constructions are similar, and when one gets the idea common to these three constructions, one may go on to construct the four-sphere, and so on, to infinity. One, two three, infinity, said Chuang Tzu. But we will be satisfied with three constructions. The first two are easy, and are described just for practice.

---

6   (Murray, 1963, pp. 90, 98)

7   (Murray, 1963, p. 91)

8   (Panofsky. 1997)

The one-sphere made of two one-balls

Let us begin in Lineland, a one-dimensional world.[9] Visualize a one-ball, that is, a line segment, say two units long. Now, double it, that is, create a copy. We have now two identical line segments. Visualize them, one on top of the other, oriented eastwest in a two-dimensional plane. Move one of the intervals away from the other, which stays unmoved. Imagine that the moving interval is pushed away in an orthogonal direction, say, to the north.

Now bend the upper (northern) segment so the bounding points are pulled downward. Next bend the lower segment so its bounding points are pulled upward. Finally, connect the corresponding bounding points, as shown in Figure 4.1. The connected line segments form a loop, or one-sphere. If we further bend the loop to make it round, we would have a perfect circle.

The two-sphere made of two two-balls

Now we are in Flatland, a two-dimensional world. Visualize a two-ball, that is, a disk, or filled-in circle, say one unit in diameter. Now double it. We have two identical disks, one on top of the other. For the next step, we need a new dimension. Imagine our plane embedded within a three-dimensional space, as a horizontal plane. Move one of the disks away from the other, which stays unmoved. Imagine that the moving disk is pushed away, upward from the other.

Now bend the bounding circle of the upper disk downward, and the bounding circle of the lower disk upward, and connect the bounding circles, as shown in Figure 4.2. The connected disks form a surface, a two-sphere. If we further bend the surface to make it round, we would have a perfect sphere.

The three-sphere made of two three-balls

Now we are in Solidland, a three-dimensional world. Visualize a ball,

---

9   See the Flatland of Abbot for more information about this world. (Abbott. 1963)

that is, a filled-in two-sphere, say one unit in diameter. Now double it. We have two identical balls, one on top of the other. For the next step, we need a new dimension. Citizens of Solidland would have difficulty with this exercise, but with imagination, anything is possible. Move one of the balls away from the other, which stays unmoved. Imagine that the moving ball is pushed away in an orthogonal direction, which we recognize as belonging to Hyperland, a four-dimensional world. If the directions of the original balls are North-South, East-West, and Up-Down, we may call the new, fourth dimension, Heaven-Earth. We have moved one of the balls Heavenward of the other.

Now bend the bounding two-sphere of the Heaven ball Earthward, and the bounding two-sphere of the Earth ball Heavenward, and connect the bounding two-spheres. The connected balls form a solid, which we are calling a three-sphere. (No figure for this one!)

Dante's cosmology

Dante was the author of numerous works, of which only one is widely known: *The Divine Comedy*. This, perhaps "the greatest poem of our tradition,"[10] is composed of three parts: *Inferno*, *Purgatorio* and *Paradiso*. Purgatory was apparently an invention of the 12th century associated with Saint Patrick, and is connected to the spheres of Hell and Paradise by a tunnel, a kind of umbilicus.[11]

In Dante's cosmology, two three-balls of the same size, bounded by two-spheres, are connected. The *Earth ball* is visualized as an onion, that is, as a family of concentric, spherical, shells. The shells are the biosphere, the atmosphere, and the lunar sphere. The bounding sphere is the celestial sphere, home of the Stars. At the center is Hell.

The *Heaven ball* is also visualized as an onion, the concentric shells are the heavenly spheres, or homes of the angels. The bounding two-sphere is again the celestial sphere. At the center is God, or Paradise.

As the two three-balls are connected by gluing together the bounding two-spheres, we recognize the three-sphere constructed in the usual

---

10 (Bergin, 1965; p. 213)

11 See Yolande de Pontfarcy (pp. 93-116) and Jean-Michel Picard (pp. 271-286) in (Barnes, 1995).

way. Other than the two exceptional points—Hell and God—the whole universe is an onion-like affair, a family of concentric two-spheres.

Dante's cosmology may have been inspired by Pseudo-Dionysius the Areopagite, who wrote of the heavenly realms of angels around 500 CE, perhaps in Alexandria.

The three-sphere full of tori

To progress from the vision of Dante to that of Fra Angelico, we must put the two three-balls together by gluing their bounding celestial spheres as before. But first, we must replace the concentric spheres of the onion structures with concentric tubes, like rigatoni. Imagine the North and South poles of each three-ball connected by a straight line segment, its polar axis. Around each of the three-balls, within their bounding (celestial) spheres, draw an equatorial circle, its equator. Each of the two three-balls now has a special line segment, its polar axis, and a special circle, its equator. Other than these two one-dimensional features, we may imagine each three-ball to be made of concentric tubes, like rigatoni, as shown in Figure 4.3.

As we glue the two bounding (celestial) spheres together, we must make sure that their North poles are glued together, their South poles are glued together, and their equators are glued together. Then the two polar axes are glued together into a one-sphere or loop, and the two equators are glued together into a loop. The rigatoni-like tubes of the Heaven ball are glued to similar tubes of the Earth ball to make tori, like inner tubes, as shown in Figure 4.4. Therefore Dante's cosmos is made up entirely of concentric inner tubes, except for the two exceptional loops. We may suggest here, as possible further evidence for this model, the design on Mary's chair in the painting. (See Figure 4.5.) Compare this with the torus of Figure 4.4, seen from above.

This vision of the three-sphere was fundamental to the creative work of the mathematician Heinz Hopf, who was one of the great pioneers of algebraic topology, a new branch of mathematics, in the 1930s.[12]

---

12 See (Dieudonne, 1989) for the somewhat technical details of this story.

Fra Angelico's angels

If indeed a time traveler suddenly appeared, we would be shocked. As hard as it would be for us to understand the culture and iconography of the future, it is equally hard to understand those of the past. A case in point is angels.[13]

Fra Angelico was an important friar, as well as one of the great painters of the Early Renaissance. As a person of great integrity, we should take him seriously when he includes angels in his paintings, despite our scepticism. His representations of Biblical scenes give the impression of sincere belief. So in Fra Angelico's two paintings (Cortona and San Marco) of the Annunciation—the announcement to Mary by the Angel Gabriel that she is to give birth to the son of God—the representation of the angelic form as a humanoid with wings must be believed. Perhaps, like Dante, he was influenced by the angelology of Pseudo-Dionysius.[14]

Fra Angelico was geometrically sophisticated for his time. This is established by his use of a formal grid in one painting, which shows a total mastery of linear perspective, newly introduced by Brunelleschi and Alberti. In fact, it is likely that Alberti himself contributed the grid to this painting.[15] However, Fra Angelico did not use linear perspective in many other paintings; he preferred a more natural style for the representation of three-dimensional space.[16] On this basis, we have argued above that the nimbus he puts around the head of his saints and angels is a three-ball bounded by a two-sphere. The perfect sphere is, after all, a divine form in the philosophy of Plato, and is the most appropriate form to serve as a symbol of the celestial nature of these transcendental beings: They belong to the Heavenly Ball of Dante's three-spherical world.

The angel wings

---

13 See (Fox and Sheldrake, 1996) for a gallant effort.

14 See (Spike, 1997, p, 64).

15 (Pope-Hennessy, 1981)

16 See several geometric analyses, in (Morichiello, 1996). See also (Spike, 1997, pp. 27, 52, 53).

Now, let us take a closer look at the shape of the wings of the angel in the Cortona Annunciation, Figure 4.5. (See also the rather similar Annunciation in the Convent of San Marco, Florence.) The probable date for this painting is 1438.[17] The wings are drawn as elliptical arcs. Recall that an ellipse is easily drawn with a loop of string and two pins. And in fact, the arc of the frontal edge of the angel's right wing matches an ellipse drawn with two foci within the frame of the picture—one of which is marked by a small nimbus! The ellipse is shown in Figure 4.6. Consider the fact that, like a halo, the boundary curves of the angel wings, seen head-on, are perfect circles.

This indicates a geometrical vision as follows. Each wing represents a circle. By stretching out the wings and flapping them, the angel carves out the tori of the cosmic three-sphere. As two spatial dimensions plus movement make three dimensions, the flight of angels may be seen as a construction of the world, and a means of navigating between the Ball of Heaven and the Ball of Earth. This prevision of the topology of the 20th Century on the part of Fra Angelico is as extraordinary as the prevision of Dante, but since the lives of both of these religious artists were informed by extensive meditation, prayer, and other spiritual exercises of visualization and active imagination, we should accept the fact that their capacity to imagine higher dimensionality might exceed our own, focused as we are on the objects and technologies of a very material world.

Conclusion

Now let us include the vision of Fra Angelico in a broader context of cultural exchange and major social transformations.[18] This transformation consists of an 800-year sequence of paradigm shifts, each one a trigger for the next like dominoes, including:

1. The cross-cultural exchanges between Islam and Christendom across the border with Byzantium, as the Crusaders bring Arabic music and poetry, Sufi mysticism, and Persian love poetry back to Iberia and Provence around 800.

---

17 See (Spike, 1997, p. 47).S

18 Following the domino theory of Flinders-Petrie, see (Abraham, 1994).

2. The partnership resurgence[19] of the Troubadours in Provence, a school of poetry brought about by this Islamic cultural exchange, with a new celebration of the spiritual qualities of women in courtly love around 1100.

3. A shift in mathematics brought to Sicily from North Africa by Leonardo of Pisa around 1200.

4. The rise of perspective in painting (Giotto), around 1300.

5. The popularity of a new humanism in literature (Dante, Petrarch, Boccaccio) around 1350.

6. The full and conscious flourescence of the Italian Renaissance (including Fra Angelico) around 1450.

7. Another shift in mathematics—the dynamic mentality of Kepler and Galileo—that consolidates the shift of attention from a static and unchanging perfect order of things to a dynamical world of objects in motion, around 1600.

There are works of art that capture the spirit of their time, but there are others that seem to leap out of the confines of their historical moment to spin into other dimensions of human presence. Fra Angelico is one of these, for he points out for our age a way of knowing that is truly integral, one in which art, mathematics, and spiritual practice can lift us beyond the application of science to the technologies of dehumanization and destruction.

---

19 See (Eisler, 1989).

FIGURE 4.la.

• We begin with a unit 1-ball, that is, a line segment two units long.

• Double the 1-ball, so there are two of them, one on top of the other. Then move one up and the other down.

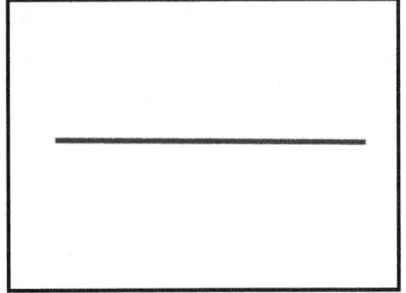

FIGURE 4.lb.

• Now we have two 1-balls, one about the other.

• Next, bend the boundary (two ends) of the upper I-ball down, and the boundary of the lower 1-ball up

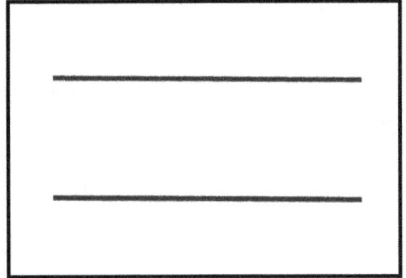

FIGURE 4.1c.

• Now we are ready for the gluing step.

• Pull the boundary (both ends) together and glue them to join smoothly.

• Now we have a loop.

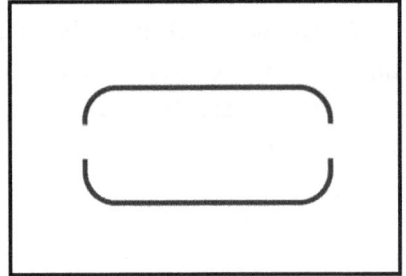

FIGURE 4. ld.

• Here is the loop.

• It is a topological 1-sphere, that is, it can be deformed into a geometrical 1-sphere.

• Go ahead. Deform it into a perfect circle.

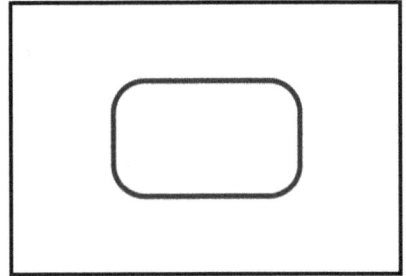

FIGURE 4.le.

• Here is the geometrical 1-sphere, a perfect circle.

• It is one-dimensional, that is, the inside is empty.

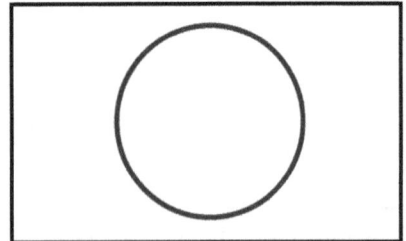

FIGURE 4.2a.

• We begin with a unit 2-ball, that is, a 2-dimensional disk, or perfect circle filled-in with a disk of flat 2-dimensional space.

• Double the 2-ball, so there are two of them, one on top of the other. Then move one up and the other down.

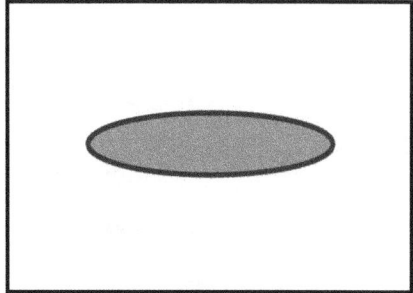

FIGURE 4.2b.

• Now we have two 2-balls, one about the other.

• Next, bend the boundary of the upper 2-ball down, and the boundary of the lower 2-ball up.

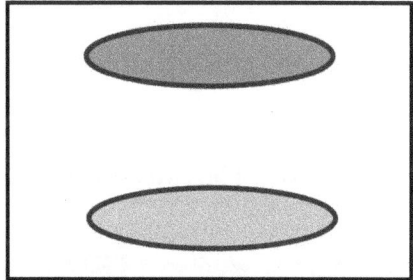

FIGURE 4.2o.

• Now we are ready for the gluing step.

• Pull the boundaries together and glue them to join smoothly.

• Now we have a cylindrical surface, like a tin can.

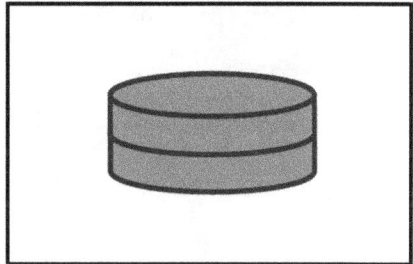

FIGURE 4.2d.

• Here is the surface.

• It is a topological 2-sphere, that is, it can be deformed into a geometrical 2-sphere.

• Go ahead. Deform it.

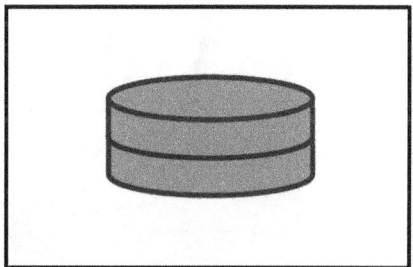

FIGURE 4.2e.

• Here is the geometrical 2-sphere, a perfectly spherical surface.

• It is two-dimensional, that is, the inside is empty.

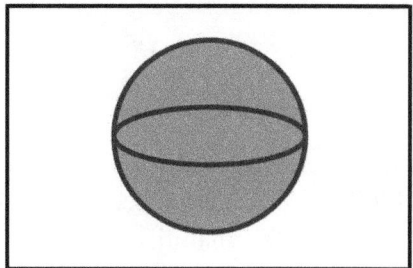

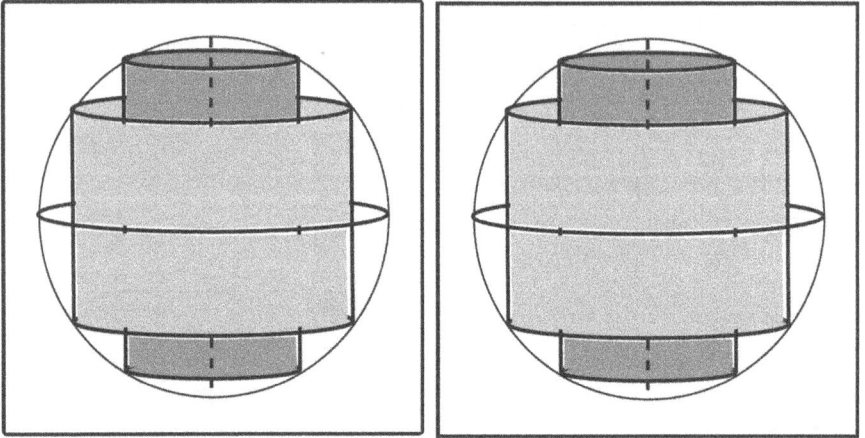

FIGURE 4.3. The two 3-balls, each filled with rigatoni, only two tubes of which are shown.

FIGURE 4.4. Two pieces of rigatoni joined to make a 2-torus, wrapped with wire for visibility.

FIGURE 4.5. The Cortona Annunciation of Fra Angelico 1438 CE. Processed from a photograph of the original in Cortona.

Figure 4.6. Ellipse superimposed around Gabriel's wings.

5. GALILEO'S FATHER

In this chapter we analyze the third of the major shifts, the A/D bifurcation from the Algebraic Mentality to the Dynamic. This coincides with the transition between the Renaissance and the Baroque periods in art history. A better understanding of the A/D shift may help us to understand the D/X bifurcation from the Dynamic Mentality to the Xaotic, in which we are now enmeshed. Vincenzo Galilei, music theorist and composer of the late Renaissance, made significant contributions to science and mathematics, generally credited to his son, Galileo Galilei. Vincenzo, this great artist and scientist of the late Renaissance, is central to our bifurcation analysis.

In the mathematical theory of catastrophic bifurcations, we frequently encounter a situation in which a new attractor appears, before an existing one vanishes. This is well known in catastrophe theory, the new branch of mathematics which began the practice of applying dynamics to cultural history, where it is called the *delay convention*. For example, the Neanderthals of the Mousterian culture of the Middle Paleolithic vanished in western Europe around 35,000 BP, long after the arrival of *Homo sapiens* from southern Africa, and the onset of the Aurignacian culture of the Upper Paleolithic epoch, around 45,000 BP. The 10,000-year delay is known as the *Middle to Upper Paleolithic Transition*, during which the two forms lived side-by-side.[1]

The A/D bifurcation

As we have seen, the Algebraic Mentality spans from (approximately) 800 to 1600 CE, and the Dynamic from 1600 to 1900. More exactly, we

---

1   (Lewis-Williams, 2002; p. 40)

will look at the third (climactic) phase of the Algebraic (A3) and the first (formative) phase of the Dynamic (D1). Following Thompson, the Literary Milestone of A3 is Dante's *Divine Comedy* (1308-1321), while those of D1 are the anonymous *The Life of Lazarillo de Tormes and of His Fortunes and Adversities* (1554), Cervante's *Don Quixote* (1605, 1615), and Descartes' *Discourse on Method* (1637).[2] We may therefore place the literary bifurcation between the end of A3, spanning roughly from 1300 to 1550, and the beginning of D1, which occupies the period 1550 to 1650 or so. Thus, the A/D shift is located in a window around 1550.

We will make a case for the experiments of Vincenzo Galilei (1520-1591) as the cusp of the mathematical bifurcation. Vincenzo, Galileo's father, began his studies of music theory around 1570. Therefore, the mathematical bifurcation occurred some 20-30 years after the literary bifurcation.

To focus more closely on this complex bifurcation event, we observe that the Renaissance period of art history is usually considered to be from 1400 to 1600, roughly the same as the climax, A3, of the Algebraic Mentality. Meanwhile, the Baroque, usually taken from 1600 to 1750, overlaps D1 and D2. So we now turn to the customary periods of Art History, and take a close look at the R/B bifurcation from the Renaissance to the Baroque around the year 1600.

The R/B bifurcation

A very nice tabulation of the art historical data is given in Arts and Ideas by the art historian, William Fleming.[3] For each period there is a chapter, for example Chapter 10 on the late Renaissance, or Chapter 11 on the early Baroque, in which the leading innovations are described in the various arts: architecture, sculpture, painting, writing, and music. In addition, each chapter contains a summary chronology of the main artists of its period. We may begin our dissection of the R/B boundary by extracting data from Fleming's chronologies of Chapter 10 and 11.[4]

---

2   (Thompson, 2004; pp. 40-41)

3   (Fleming, 1968)

4   (Fleming, 1968; pp. 264, 292)

The R/B shift for painting.

Studying Fleming's data for painting, we find these terminal data:

- 1483-1520, Rafael Sanzio (last of the Renaissance)
- 1430-1516, Gentile Bellini (first of the Baroque)

As the beginning of Baroque painting is given as 1496, we may consider the interval [1496, 1520] as the delay of the actual R/B transition in painting, according to Fleming.

But we may refine this further. Considering the analysis of Walter Friedlaender on this time frame, another style intervenes, Mannerism.[5] Thus we must consider a complex bifurcation event, R/M/B, in place of the simple R/B shift we have been trying to place on a time line. Following Friedlaender, the High Renaissance gave way to Anticlassicism (the first phase of Mannerism) around 1515 or 1520. This changed to Anti-mannerism (the second phase of Mannerism) around 1580 or 1590. Thus, we may bracket the R/M bifurcation event in the window 1515 to 1523, overlapping High Renaissance painting until 1520, and Mannerism fully formed by 1520. And he places the beginning of Baroque painting around 1600. We may consider the transition interval or the R/M complex event as [1515, 1600], instead of the shorter interval [1496, 1520] from Fleming's data.

The R/B shift for architecture.

Similarly, we find from Fleming, for architecture:

- 1556 – 1629, Carlo Maderno (last of the Renaissance)
- 1486 – 1586, Jacopo Sansovino (first of the Baroque)

Note that the Baroque begins, in architecture, about a century before the Renaissance ends. This is to be expected, following the delay convention. Reading further in Fleming's Chapter 11, we find the

---

5   (Friedlaender, 1990)

beginning of Baroque architecture in 1536, with Sansovino's Library of St. Mark in Venice. We may therefore consider the interval [1536, 1629] as the delay of the actual R/B transition in architecture.

The R/B shift for music.

Finally, examining Fleming's data for music, we find,

• 1532-1594, Orlando di Lasso (last of the Renaissance)
• 1480-1562, Adriano Willaert (first of the Baroque)

The beginning of Baroque music is given as 1527 by Fleming. For music, the Baroque era may be divided into three phases, early, middle, and late. In Italy, where it all began, the dates have been given as:

• Early Baroque, 1580-1630
• Middle Baroque, 1630-1680
• Late Baroque, 1680-1730

Dates in other countries may be a decade or two later.[6]

Combining Fleming for the Renaissance and Bukofzer for the Baroque, we may consider the interval [1580, 1594] as an estimate of the delay of the R/B transition in music. We will return to this estimate later for further refinement.

As indicated above, we propose to establish the timing of the mathematical A/D bifurcation as a bolt from the blue, that is, a significant contribution to mathematics by an artist. In this case, the mathematical shift is to be credited to Vincenzo Galilei, a musician, so we seek to further refine the delay interval of the R/B bifurcation in music. For this we may look to the specialists of musicology: *Music in the Renaissance*, the 1020-page magisterial work of the American musicologist, Gustave Reese (1899-1977), *Music in the Baroque Era – From Monteverdi to*

---

6   (Bukofzer, 1947; p. 17)

Bach, the 490-page classic of the German-American musicologist Man-fred Bukofzer (1910-1955), and *Baroque Music*, a brief but panoramic survey by Claude V. Palisca. In these works we find,

- 1574-1638, John Wilbye (last of Northern Renaissance)[7]
- 1557-1612, Giovanni Gabrieli (first of Italian Baroque)[8]

The beginning of Baroque music is given as 1597, and the end of the Renaissance as 1637. We might thus consider the interval [1597, 1613] as the delay of the actual R/B transition in music. But much credit is given to Ciprano de Rore as the first pioneer of the shift, by his contemporary Giovanni de' Bardi. Palisca refers to Cipriano as a Mannerist, and his pioneering anti-classical madrigal is dated 1557.[9] Hence we will take [1557, 1637] as our transitional interval for music.

Summary

Summarizing these determinations, we have these transition intervals,

- 1496-1520, painting
- 1536-1629, architecture
- 1557-1613, music

Our goal in this article is to add mathematics to this list. This will require a diversion into ancient Greek music theory, in order to understand the theoretical work of Vincenzo.

Ancient Greek Music Theory

We have evidence of a high level of music theory among upper pa-leolithic cultures.[10] Nevertheless, the sophistication of the music theory

---

7   (Reese, 1959; p. 828)

8   (Bukofzer, 1947; p. 21)

9   (Palisca, 1968; p. 13)

10 eg, cf my MS 82, the Canon of Lespugues

known to Pythagoras and his successors in ancient Greece is nothing short of astonishing. Here we will review the minimum that we will need, the numerology of the musical scales.

Measuring the Scale

In order to describe the musical scale, we are going to use the modern (since 1930) description of the pitch of a note in terms of the frequency of vibration of the sound wave of the note, measured in Hz, Hertz, or cycles per second. Eventually we will also measure pitch in units of relative string length, as has been the human habit since paleolithic times. There is a reciprocal relationship between these two descriptions. That is, if two pitches have a frequency ratio of A:B, then they have a string length ratio of the reciprocal, B:A.

For convenience, let us think of the white notes of the piano keyboard as a measuring stick. The standard keyboard includes 88 keys, comprising seven octaves plus three extra notes. Let us consider each C key as an index mark on this ruler. The fourth C key from the bottom (left) end is called middle C, and sounds the note of frequency 261.626 Hz in the standard tuning. We denote this key and note as C4. The C an octave above is C5, and so on. Thus on our ruler we have marked C1, C2, C3, C4, C5, C6, C7, and C8, which is the highest note of the piano, at the right end of the keyboard.

Pythagorean Harmonics

There are two fundamentals of the Pythagorean scale. The first is the harmonic series as an aspect of human physiology. The second is the sacredness of rational numbers, that is, ratios of natural numbers. We will come to this later. First then, the harmonic series. In music theory an interval comprises two notes played or sung together. In case the two notes are of the same pitch, the interval is called a *unison*. We say the notes are in the ratio, 1:1. In case one pitch has twice the frequency of the other, the interval is an *octave*. The frequencies of the notes are in the ratio, 2:1.

Similarly, a perfect (or Pythagorean) *fifth*, for example C4 plus G4, has frequency ratio 3:2, and equivalently, string length ratio 2:3. The perfect *fourth*, for example C4 plus F4, has frequency ratio 4:3, and length ratio 3:4. Note that a perfect fifth above a perfect fourth is an octave, as (3:2) * (4:3) is (2:1). Likewise a fourth above a fifth is an octave. These four intervals – unison, octave, fifth, and fourth – are the primary determinants of the *Pythagorean scale*.

But beware, our modern piano is not tuned to the Pythagorean scale, but rather to an approximation to it called *equal temperament.* Nevertheless, the 50 white keys from C1 to C8 provide a useful ruler and guide for this discussion.

The Cycle of Fifths

From the human voice we may observe the harmonic series, also known as the cycle of fifths. Let us begin at the bottom of our ruler, C1. Moving upward in one-octave jumps, we reach C8 in seven jumps. Now let us begin again at C1, and move upwards in jumps of a Pythagorean fifth. This is the overtone or harmonic series.

After twelve jumps, we arrive near to C8, but there is an error. Seven octaves is a frequency ratio of 2 to the power 7, or 128, to 1, while twelve Pythagorean fifths is a frequency ratio of (3/2) to the power 7, or 129.746337890625 to 1. The error, perhaps inaudible for most listeners, is is the ratio 129.746... to 128, approximately 74/73. This is called the *Pythagorean comma.*

The Pythagorean Ditonic Scale

Middle C, C4, and the G above it, G4, comprise a Pythagorean fifth, This G and the D above, D5, is again a fifth. We want D5 to belong to our scale, but it is bit high, so an octave below it, D4, should also belong to our scale. But this is the process that generates the Pythagorean scale. The ratio of the frequency of D4 to that of C4 is (3/2) * (3/2)/2 or 9:8, an interval called the *major tone*. We now have the beginning of he Pythagorean scale, C4, D4, ... , G4, ..., C5. The interval G4 to C5 is the

Pythagorean fourth, and has the ratio $2/(3/2) = 4/3$ to 1, or 4:3. Thus a fifth followed by a fourth is an octave. The fourth above the tonic, C4, defines F4. So now we have, C4, D4, ..., F4, G4, ..., C5. Note that the interval F4 to G4 has the frequency ratio, $3/2/4/3=9/8$, again the major tone. If we try to make the next note, E4, a major tone above D4, we obtain the interval $(9/8)2$ or $81/64$ to 1. This is Pythagorean major *third*. Completing the Pythagorean scale requires more musical arithmetic similar the above, and may be found in many texts.[11]

2.5 The Ptolemaic Syntonic Scale

We come now to the second basic principle of Pythagoras, the sacredness of rational numbers, that is, ratios of natural numbers. We want the intervals of our scale to be described by ratios of integers, the smaller the better. The fifth has ratio 3:2, the fourth has 4:3, and we would like the third (the interval corresponding to two tones) to have the frequency ratio 5:4. Note that $81/64 = 1.265625$ which is close to 5/4, but not close enough: it sounds dissonant.

Consider a slightly smaller tone, the *minor tone*, of ratio 10:9, Then the third obtained by a major tone followed by a minor tone would have a ratio of $(9/8) * (10/9) = 10/8=5/4$, as desired. This interval is called a just major third, and defines the note E4 of the syntonic scale of Ptolemy. The difference between two major tones and a major third is the ratio $(81/64)/(5/4) = 324/320 = 81/80 = 1.0125$, called the *syntonic comma*.

Note that the major triad so important for all polyphonic music from the Renaissance to the present – for example, C4, E4, G4 – is not obtained from two major thirds combined, for $(5/4) * (5/4) = 25/16 = 1.5625$, while the Pythagorean fifth is $3/2 = 1.5$. Instead we observe that the interval from E4 to G4 is 6:5, the just minor third. Thus the just major third followed by the just minor third is the Pythagorean fifth, as $(5/4) * (6/5) = 6/4=3/2$.

The syntonic scale is filled out by the addition of the just major *sixth*, 5:3, and the just minor sixth, 8:5. All this is difficult, and has required

---

11 We especially recommend (Cohen, 1984; Sec. 2.2).

much attention from music theorists from the beginning of polyphonic music to the present. But now the hard part is over.

Renaissance Music Theory

So the Baroque began in the late 1500s, followed by the end of the Renaissance a few years later, and this delay, or transition interval, occurred sequentially in various cultural layers. For example, in the arts, the Baroque arrived first in painting, then architecture, and finally in music. After our excursion into musical arithmetic, we now return to music in the late Renaissance and early Baroque.

Summary of Renaissance Music

Here is a partial lineage of Renaissance music masters.

- Guillaume Dufay (1397-1474), Franco-Flemish,
- Johannes Ockeghem (1410-1497), Franco-Flemish,
- Josquin des Prez (1445-1521), Rome, emulated by,
- Jean Mouton (1459-1522), Paris, teacher of,
- Adrian Willaert (1480-1562), Venice, teacher of,
- Cipriano de Rore (1516-1565), Franco-Flemish,
- Gioseffo Zarlino (1517-1590), Venice, teacher of,
- Vincenzo Galilei (1520-1591), father of Galileo Galilei.

Dufay, an early master of *polyphony*, wrote masses, motets (including isorhythmic motets), hymns, and other typs of sacred music, and also chansons of secular music. Josquin was a master of motets, on account of which is regarded among the greatest composers of the Renaissance. Willaert is known primarily for his Italian madrigals and French chansons. He was the teacher of Cipriano de Rore, who succeeded him as music master at St. Marks, and Zarlino, leading up to the R/B bifurcation. Cipriano de Rore was noted for his madrigals and motets, and as pioneer of a new style. Zarlino, noted for his motets as well as the book discussed below, and Vincenzo, are main characters in our story.

Changes of Style

It is time to consider the characteristics of musical style in the late Renaissance, and how they changed into those of the Baroque. This is truly the domain of experts, and I will rely on Bukofzer, who has given us a concise summary list.[12] Here are the ten changes of his list.

1. A single Renaissance style becomes several in the Baroque.
2. The restrained representation of words becomes affective.
3. Balanced voices yield to polarity of the outermost voices.
4. Melody from small diatonic range to wide chromatic range.
5. Counterpoint changes from modal to tonal.
6. Harmony, dissonance change from intervallic to chordal.
7. Chords change from accidents to self-contained entities.
8. Chordal progressions from modality to tonality.
9. Rhythm from evenly flowing to extreme pulsations.
10. Idioms of voice and instrument from same to different.

These changes are explained by Bukofzer in a chapter of 19 pages. Regarding the second, credit has been given to an an informal group of artists in Florence, founded by Giovanni de' Bardi (1534-1612), Count of Vernio, called the *Camerata*.[13] Their influence was strongest from 1577 to 1582, so we must count this among the earliest steps into Baroque music. But we will be concerned with only one of these ten elements, the sixth: the treatment of *dissonance*.

What, in fact, was regarded as dissonance in the sixteenth century? As we have seen in the preceding section on musical arithmetic, during the dominance of the Pythagorean scale from Ancient Greece through the Middle Ages, only four intervals were considered consonant: the unison, octave, fourth, and fifth. Thirds and sixths, in this scale, were audibly harsh. As long as musical performance was monophonic, and intervals were sequential (intervallic) and not simultaneous (chordal), these dissonances were not heard. Early polyphony, from around the year 1000, simply avoided the dissonant intervals. But with the advent of

12 (Bukofzer, 1947; p. 16)

13 (Bukofzer, 1947; p. 5)

more complex polyphony in the Renaissance, the syntonic scale became increasingly attractive.

Cipriano de Rore

Regarding the sixth change of style, the sixteenth century practice exemplified by Willeart severly restricted dissonant intervals, such as seconds and sevenths.[14] Late in this century, composers began taking more liscence with these intervals. Monteverdi, one of these pioneers, called this the *second practice*, reserving the first practice for the rules given by Zarlino in his book, *The Harmonic Institutions* of 1558. Cipriano was among the earliest to follow this second practice. While following first practice in his compositions for the church, he was more experimental in his secular madrigals. One example, published in 1557, was expecially praised for its novelty by the younger artist, Giovanni de' Bardi (1534-1612). This was novel not only in the treatment of dissonance, but also in the adaptation of the melody to the rhythm of the poetic lyrics, as in the second change listed above.

Zarlino

Willaert was the choirmaster of Saint Mark's Cathedral in Venice from 1527 until his death in 1562. He was succeeded by his pupil Cipriano de Rore, who resigned in 1565, and then by Zarlino, who served until his death in 1590. In addition to composing and directing, Zarlino is considered the greatest music theorist since Aristoxenus. His main work, the four-part *Le Istitutioni harmoniche* (The fundamentals of harmonics) of 1558, was written while Willaert was still choirmaster at Saint Mark's, and Zarlino and Cipriano de Rore were his students. Part III, *The Art of Counterpoint*, appeared in English for the first time in 1968, in the translation of Guy A. Marco and Claude V. Palisca. We rely here on the 14-page introduction by Palisca.[15]

Parts I and II of Zarlino's *The fundamentals of harmonics* are theoretical, while Parts III and IV are practical. Part I is a course on musical

---

14 (Palisca, 1968; p; 9)

15 In (Zarlino, 1558/1968; pp. ix-xxii). See also (Reese1959; p. 377).

mathematics, the outline of which we have presented in the preceding section. Part II deals with the Greek tonal system, Greek music in general, the theory of consonances, and a system of tones based on Renaissance Neoplatonism. Part III is a practical course in composition. The music theory in this classic text, derived in part from the teaching of Willaert, is still studied today. Part IV is devoted to the Greek modes. Harmony, the theory and practice of consonances, is treated throughout the four parts. Zarlino believed that music is the harmonizing agent of the universe, and through proportions, coordinates the heavenly spheres, holds the four elements together, and organizes time and the seasons.[16] As far as tuning is concerned, Zarlino favored the syntonic scale of Ptolemy.

The Baroque

The Renaissance/Baroque (R/B) boundary in music may be placed between Zarlino and Vincenzo. A crucial role was placed by the Camerata.

The Florentine Camerata

Count Giovanni de' Bardi was a leading intellectual of Florence. Composer, poet, theater director, and entrepreneur, he was the instigator and host of the Camerata. This group – including musicians (Vicenzo Galilei, Giulio Caccini), poets (Ottavio Rinucchini, Giovanni Guarini), and scientists – met at Bardi's home from around 1570 to 1592. Under the influence of letters from Girolamo Mei (1519-1594) in Rome (especially that of 1572 to Vincenzo) the group became interested in reviving ancient Greek music.[17] Mei believed that only monody (music having a single melodic line), not polyphony, could move the emotions of the listener. This interest led to experiments in monody by composers such as Vincenzo, Caccini, Peri, Corsi, and others, triggering the R/B shift in music, including the creation of opera as an art form, Caccini's *L'Euridice* of 1600 being among the first.[18]

---

16 (Zarlino, 1558/1968; p. xiv)

17 (Palisca, 1989; Ch. 3, pp. 45-55)

18 For this history we are indebted to (Palisca, 1989; Ch.1, pp. 1-12)

Vincenzo

Born near Florence, Vincenzo mastered the lute at an early age. He composed a number of madrigals, two-part counterpoints, and other arrangements. From 1563 to 1565 he studied music theory with Zarlino in Venice, under the patronage of Giovanni de' Bardi. Soon after, in connection with Bardi, Mei, and the Camerata, he became interested in ancient Greek music theory. Through Mei's inflluence, Vincenzo became disillusioned with Zarlino's theory. In 1581 he published *Dialogue on Ancient and Modern Music*, the second most influential treatise on music theory of the Renaissance.[19] This book, dedicated to Bardi, was largely a critique of Zarlino. It rejects modern polyphony and extols ancient Greek monophonic songs. It is ambivalent on the virtues of the Ptolemaic syntonic scale.

The *Dialogue* of 1581 reveals Vincenzo as a pioneer of physical acoustics, and of experimental psychology. There he reported on two experiments. First was a project of physical experiments on the effect upon pitch of the dimensions of organ pipes. Second was an experimental test of the aesthetics of the Pythagorean ditonic scale, in comparison with the Ptolemaic syntonic scale. This test supported the empiricism of Aristoxenus of Tarentum, who had advocated (around 335 BCE) experience rather than theory as the best guide to tuning musical instruments. A similar approach was advocated by Giovanni Benedetti (1530-1590) in 1585.[20] In a second book, the *Discorso* of 1589, Vincenzo calls experiment *the teacher of all things*.[21] In this empiricism, he anticipates the work of his son, Galileo, just a few years later.

Galileo

Galileo (1564-1642) is very well known as the father of the Scientific Revolution, and the Dynamic Mentality. In fact, William Irwin Thomp-

---

19 For a careful synopsis, see the 32-page Introduction by Palisca in (Galilei, 1581/2003).

20 (Cohen, 1984: Sec. 3.1)

21 Palisca, in (Galilei, 1581/2003; p. xxviii).

son calls this the Galilean Dynamical Mentality. Ignoring Galileo's many other accomplishments—such as the first telescopic observations of the solar system, confirming the heliocentric model of Copernicus, the atomic theory of the secondary attributes of matter, and so on—we will focus here on his earliest work on dynamics, which has been caricatured as dropping cannon balls from the leaning tower of Pisa.

Galileo was born in Pisa, enrolled in the University of Pisa to study medicine, but then changed to mathematics. In 1589 he was appointed professor of mathematics in Pisa. In his book *Discorsi* of 1638 he expressed the kinematic law, that a body falling with uniform acceleration, independent of its mass, with the distance proportional to the square of the time elapsed. But as early as 1589, as a junior professor in Pisa, he had established this law experimentally, using apparatus still on display in the Galileo Museum in Florence. This brought on the hatred of his Aristotelian colleagues, who forced him to flee from Pisa, in 1591, to the University of Padua.[22]

On this account, we may place the birth of the Dynamic Mentality, and the A/D bifurcation, in the year 1590. There is little doubt that Galileo's empirical approach to kinematics around 1590 was derived from his father's experiments in musical aesthetics and pitch, which took place (perhaps at home in the kitchen) between 1572 and 1581, coinciding with the birth of Baroque music around 1580.

Kepler

Galileo (1564-1642) never left Italy, and Kepler (1571-1630) never visited there, so the two never met. They did exchange a few letters, but somehow, their ideas and accomplishments are closely related. Kepler's works span many fields and include several books, including the *Mysterium Cosmographicum* of 1596, containing his model of the solar system based on the five Platonic solids, the *Astronomia Nova* of 1609, with his first law (elliptic orbits) and second law (equal areas in equal times, the first recorded example of an ordinary differential equation) of planetary motion, and *Harmonice Mundi* of 1619, containing the

---

22 (Ridondi, 1987)

third of his laws (the period-distance relation). We may regard the publication the *Astronomia Nova* in 1609 as the culmination of the A/D shift in mathematics.

The connection between Kepler and our story of the R/B shift of 1580 and the A/D bifurcation of 1590 includes this story. Kepler's mother was accused of witchcraft in 1615. En route in a carriage to testify in her defense, Kepler read Vincenzo's *Dialogue on Ancient and Modern Music* of 1581, from which he developed his ideas for the harmony of the spheres, which found experession in his *Harmonice Mundi* of 1619.

Conclusion

This long story on the complex bifurcation events approaching 1600 may be summarized in this chronology.

- 1496-1520, R/M/B shift for painting
- 1536-1629, R/B shift for architecture
- 1580-1613, R/B shift for music
- 1589-1609, A/D shift for math

Regarding the R/B shift for music, leading up to the A/D shift, we have this finer chronology.

- 1527, Zarlino appointed in Venice
- 1557, Cipriano de Rore, the second practice
- 1558, Zarlino publ. *The Fundamentals of Harmonics*
- 1563, Vincenzo studied with Zarlino
- 1572, Camerata meetings begin
- 1572, Mei's letter to Vincenzo
- 1572, Vincenzo's experiments, tuning, organ pipe pitch
- 1580, the Camerata trigger the shift to Baroque music
- 1581, Vincenzo, *Dialogue on Ancient and Modern Music*
- 1589, Galileo triggers the Dynamic Mentality

In sum then, we have a series of shifts from 1500 to 1600: painting, architecture, music, mathematics, mentality. Further research could add sculpture (eg, Michelangelo), poetry (eg, John Donne), and philosophy (eg, Leibniz) to this cascade.

This chronology suggests further questions: What triggered the first shift? How would a shift propagate from painting to sculpture, etc? Were alchemy and psychoactive substances involved in the R/B shift?

6. THE FRACTALS OF PARIS

The fourth major shift, the D/X bifurcation, is ongoing at present. This is, of course, the only one of which we have personal, living, experience. Our analysis of the preceding shifts may help us to comprehend this one.

Introduction

Chaos theory and fractal geometry came to public attention only after the arrival of computer graphics in the 1970s permitted us to see the fascinating patterns of chaos and fractals. But chaos theory itself began almost a century before, with the mathematical discovery of the first chaotic object. This was the homoclinic tangle in the 3-body problem of celestial mechanics, discovered by Henri Poincaré (1854-1912) in Paris, in December of 1889.[1] Almost 30 years later, Gaston Julia (1893-1978) also in Paris, found Poincaré's chaotic object in a much simpler context. Recent research showed that this second chaotic object, now known as the Julia set, was a fractal.

The homoclinic tangle first became visible in the analog computer graphics of Yoshisuke Ueda, then an electrical engineering graduate student in Kyoto. He was the first to view a chaotic attractor, in 1961. And the Julia set emerged into view around 1977, in the digital computer graphics of the late Benoit Mandelbrot, the mathematician who coined the word fractal around 1975.

At the time of Poincaré, Frantisek Kupka (1871-1957), a Czech painter, was also working in Paris. Kupka painted images very similar to fractals as early as 1910 —truly a bolt from the blue! During this period,

---

1   (Barrow-Greene, 1997, p. 69)

chaos also appeared in the musical compositions of Erik Satie (1866-1925), who lived and worked in Paris.

Poincaré, Chaos, and Fractals

Jules Henri Poincaré earned the Ph.D. in mathematics from the Ecole Polytechnique in Paris in 1879. His talent was quickly manifest, and he became Professor at the University of Paris in 1881. He is now considered one of the great mathematicians of all time for his contributions in several traditional branches of mathematics, as well as his pioneering work in the creation of algebraic topology.

Also to Poincaré's credit is the discovery of chaotic dynamics in 1889, as he discovered that the solar system was not a smoothly rotating system, but a chaotic one. He attempted and failed to prove the stability (that is, perfect periodicity) of the restricted three-body problem (one of the simplest mathematical models for the Sun-Earth-Moon system, based on Galilean dynamics), a prize problem posed by King Oscar II of Norway and Sweden. The obstacle was the chaotic object called the homoclinic tangle. This shift in the perception of our solar system is as profound for us as the Copemican/Keplerian one had been for the Renaissance world view.

Poincaré wrote over 500 papers and several books, including four nontechnical works, before his premature death at the age of 58. At the beginning of the twentieth century his four popular books were widely read by artists and intellectuals in France, and had an effect on the development of the modem art movement in Paris.[2]

- *Science and Hypothesis* (1902),
- *The Value of Science* (1904),
- *Science and Method* (1908), and
- *Last Thoughts* (1913).

English translations of the first three of these were republished in one volume, *The Foundations of Science*, in 1913. Among the

2   (Henderson, 1983, pp. 36, 97) (Shlain, 1991, pp. 195, 431)

influential ideas are: noneuclidean geometry, the fourth dimension, x-rays, spatial perception, and chance and probability.

From our perspective today, Poincaré's discovery in December of 1889 of the homoclinic tangle is the first example of a chaotic dynamical system.[3] About this object, he wrote in Volume 3 of his technical text, *Methodes Nouvelles* of 1892:

> One must be struck by the complexity of this shape, which 1 do not even attempt to illustrate. Nothing can give us a better idea of the complication of the three-body problem, and in general of all problems of dynamics for which there is no uniform integral.[4]

Almost every problem of dynamics has this complexity, or chaotic behavior. The first drawing of the homoclinic tangle was made by the American mathematicians George Birkhoff and Paul Smith in the 1930s.

Laplace had thought that if we knew the exact configuration of the solar system at one moment, we could predict its future configurations forever. This idea, known as *Laplacean determinism*, became one of the main tenets of modern science. The full significance of Poincaré's discovery became widely appreciated only recently: *Laplacean determinism is doomed.*[5] But Poincaré certainly understood this. In *Science and Method* he wrote:

> A very small cause that escapes our notice determines a considerable effect that we cannot fail to see. and then we say that the effect is due to chance.[6]

The revolutionary artists of the modernist movement were greatly stimulated by the paradigm shifts of the sciences, and these new ideas of

---

3   (Barrow-Greene, 1997, p. 69)

4   (Mandelbrot, 1983, p. 414)

5   (Smale, 1980)

6   In Chapter 4 of Part I, titled "Chance". See (Peterson, 1993, p. 167).

space and time were presented to Duchamps[7] and Picasso by aficionados of the avant-garde in the cafe Le Lapin Agile.[8] In England and America, popular accounts by Bertrand Russell and Arthur Eddington were also widely influential.[9]

In the wake of Poincaré's discovery, chaos theory and fractal geometry gradually took shape. But from all this history, we wish now to pull out just one thread: the Julia set.

Julia

Paul Montel, professor of mathematics in Paris, was a contemporary of Poincaré. Among his students was Gaston Julia (1893-1978). Julia took up a problem left unresolved by Poincaré's death in 1912, and discovered the homoclinic tangle in a much simpler context (a quadratic map from the plane to itself) than that of Poincaré (the restricted three-body problem of celestial mechanics). George Birkhoff in the US pursued a similar line of research. In 1918, at age 25, Julia published a very long paper on his tangle, later known as the Julia set, which won him instant mathematical fame. His version of the complicated set, like Poincaré's, was very difficult to visualize. In 1925, the slow process of visualization of the Julia set began, in a simple sketch by Cremer, a mathematician in Berlin.[10] This process languished until computer graphics came into the hands of Benoit Mandelbrot.

Mandelbrot

Benoit Mandelbrot (1925-2010) was a student at the University of Paris and studied under Julia. Beginning around 1967, he formalized

---

7   The influence of Einstein and Poincaré on Cubism is a topic of much recent discussion among art historians. See (Henderson, 1983) and (Shlain, I99I,; Ch. 15).

8   A wonderful presentation of the influence of the new mathematics and the new science on art is to be found in (Miller, 2001). See page 14 for a discussion of the cafe Le Lapin Agile.

9   (Waddington, 1969/1970, p. 100)

10   (Peitgen, Jurgens, and Saupe, 1992, pp. 138-139)

earlier work on the strange sets now known as fractals, and named and created *fractal geometry*. This subject may be viewed like noneuclidean geometry, as a radical extension of the classical geometry of ancient Greece. Considering Plato's command, "Let no one ignorant of geometry enter the academy", and Galileo's dictum that the book of nature was written in the language of geometry, the discovery of fractal geometry may be seen as the opening of a new code book for the whole of science. In fact, Mandelbrot's first book on the subject is titled, *The Fractal Geometry of Nature*.

Around 1979, Mandelbrot created very detailed images of the Julia set, for many different examples of the family of plane mappings studied by Julia. He recognized it as a fractal. It belongs to the small overlap of chaos theory and fractal geometry, two new branches of mathematics owing much to computer graphics. These fractal images from chaos theory are the best known fractals, although they represent only a very regular kind of fractal. The Julia set came into popular consciousness after 1980. However, by 1910, it was already well known to one person, Frantisek Kupka.

Kupka

Frantisek Kupka (1871-1957) was bom in Eastem Bohemia, the eldest of five children. He learned drawing from his father, but his early drawings were destroyed by his stepmother. By age 27 he was apprenticed to a saddler who also worked as a spiritualist. Kupka developed a talent for leading séances, and supported himself as a medium after becoming a painter a few years later. He studied painting in Prague, Vienna, and Paris, and began exhibiting in these cities in 1899. Of note:

- 1892, Kupka encountered theosophy, became vegetarian.
- 1895, Kupka read extensively in philosophy, including Kant, Shopenhauer, Bergson, and Nietzsche.
- 1905, Kupka's exhibition a great success. He began attending lectures in physics, biology, and physiology at the Sorbonne, and worked in the biology laboratory there.

- 1910, Kupka destroyed most of his paintings, and began a new phase. This is the year of Kupka's first abstract paintings, including the Amorpha series.
- 1911, Kupka attended meetings with a group of artists and writers every Sunday morning at the home of his neighbor, Jacques Villon.
- 1912, Kupka continues to attend the meetings. A mathematician, Maurice Princet, also attended. Kupka's first abstract paintings were exhibited in Paris.

We now see the first precognition of fractal and chaotic images in the Amorpha series of 1910-1912.[11]

Satie

Erik Satie (1866-1925) was on the forefront of the Parisian musical scene at the time of Poincaré's discovery of chaotic dynamics. He was also involved with Sar Péledan, the high-priest of the Rosicrucian and Chaldean Confratemity, an esoteric group. He composed music for the group in 1890. By 1900, he was incorporating American Ragtime styles in his compositions, an early form of jazz. Even in the Paris of 1905 he was an eccentric composer and legendary figure of cafe society.

Sergei Diaghilev (1872-1929) directed the Ballets Russes, a ballet company that was very popular in Europe before the First World War. Jean Cocteau (1891-1963)—French poet, dramatist, and leader of the literary avant-garde, was a close friend of Diaghilev. In 1915, Cocteau was commissioned by Diaghilev, then in Rome, to write a ballet. The idea for the ballet came to Cocteau in April of 1915, during a holiday from the war front, after hearing a piano duet by Satie and Ricardo Vines. Cocteau enlisted Léonide Massiné (1895-1979), a daring Russian choreographer, along with Pablo Picasso and Satie to work as a team on this project. Work began in the Spring of 1915, The scenario by Cocteau, choreography by Massiné, and sets and costumes by Picasso, were created in Rome, while Satie began working in the spring of 1916 on the music at a cafe in Paris.

---

11 See (Andel, 1977. p. 85), (Kosinski, 1998), and (Waddington, 1970, p. 22).

The result was the surrealist ballet *Parade*, which was produced at the Théatre du Chatelet in Paris on May 18, 1917, just as the Dada movement was getting underway. Revolutionary and shocking, *Parade* manifested Cubism in music, choreography, sets, and in costumes. The audience rioted. Critics called the musical score—that called for revolvers, sirens, clappers, typewriters, lottery-wheels, Morse code keys, and so on—a "mere din." Appolonaire's review of the ballet coined the word *surrealism* to describe this avant garde revolutionary ballet.

Progressing from order (an introductory fugue) through syncopation (inspired by ragtime) to chaos (a frenzied dance finale), the musical score suggests to us now the bifurcation diagram of the logistic function basic to chaos theory, which progresses from simple rhythms to more and more complexity in a sequence of subtle bifurcations converging finally to a chaotic form. The score also suggests the fractals of Julia as first shown in computer graphics by Mandelbrot.

At this point, we may confidently credit the team of Cocteau, Massiné, Picasso, and Satie with an authentic bolt from the blue, in its anticipation of chaos theory, stimulated only by Poincaré's books But to cap it all, this artistic team worked together again to produce the ballet *Mercure*, first performed in Paris in 1924, after the Dada movement had been superseded by surrealism. This ballet included twelve principal scenes, one of which is actually called Chaos. Satie's score for this scene combined the themes of two earlier (and very disparate) scenes to suggest chaos with the coupling of two electronic oscillators—which is surprisingly similar to the discovery of chaotic behavior in electronic circuits by Van der Pol, the pioneer of early radio technology working in Holland at this same time.[12]

Conclusion

We may now sit back and compare the homoclinic tangle of Poincaré, as seen in 1961 (Figure 6.1) and the Amorpha painting of Kupka of 1912 (Figure 6.2). Many more such comparisons could be made. Setting aside paranormal explanations, séances, and the like, we know

---

12 See (Templier, 1969, pp, 36-41), (Harding, 1975, p. 155), and Myers, 1947, pp. 49, 102-105).

of only these connections from the Poincaré line to Kupka and the
Satie team:

- Poincaré's first two popular books appeared in the original French
  editions in 1902 and 1904, and were extensively read by Parisian
  intellectuals, including the painter Duchamps.
- Kupka spent most of 1905 reading philosophy books, and
  attending lectures and working at the Sorbonne, where Poincaré
  was then lecturing.
- Cocteau and the Satie team were intimate with the literary and
  artistic elite of Paris from 1905 on.

We are left with no choice but to award two bolt-from-the blue
medals for the anticipation of chaos theory, one to Kupka, and the
other to the Cocteau and Satie team. Again, note that the direct influence of math and science on art does not merit a bolt-from-the blue
medal—only the anticipation of new discoveries in math and science
by an artist counts here.

For Kupka, we may award 10 on a scale of 10 for boltsmanship, for
his images of the Julia set, and also of the homoclinic tangle of Poincaré
as drawn by Birkhoff and Smith in the 1930s, are shockingly prescient.
For Team Cocteau we propose a score of 5 on the bolt scale, as they made
use of the idea of Poincaré that Laplacean determinism had come to an
end in a way typical of Cubism, Dada, and Surrealism, and were only
precognitive in identifying it with the label of chaos, in anticipation of
the mathematical developments of the 1970s,

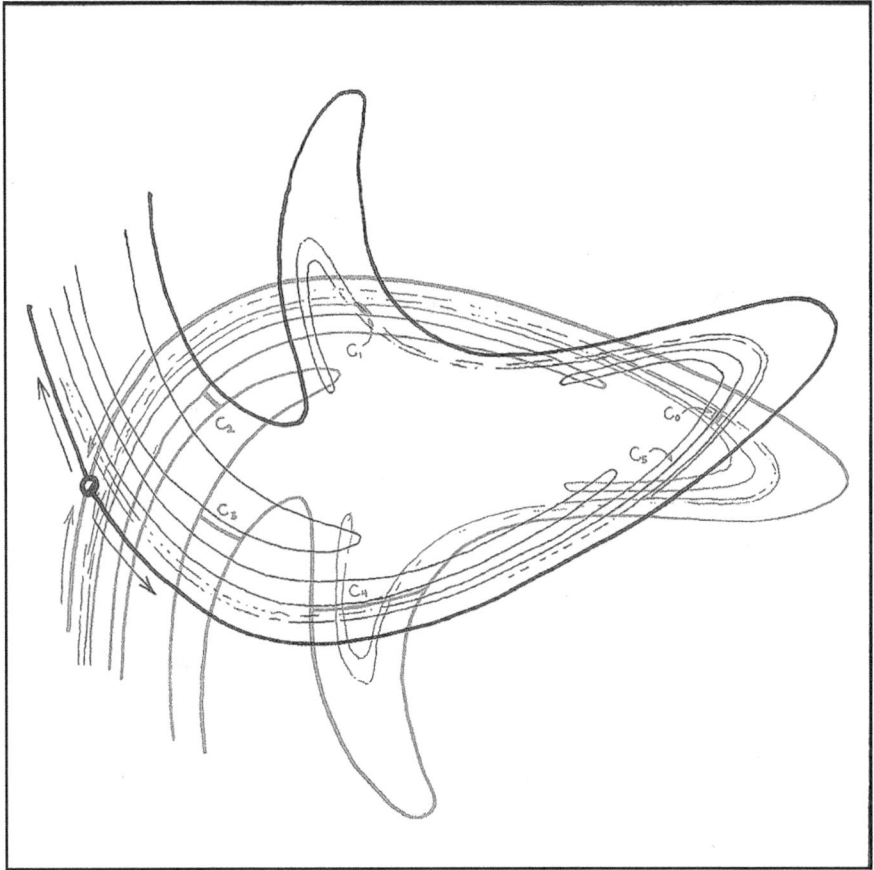

Figure 6.1. The homoclinic tangle of Poincaré. Drawing by Y. Ueda, from Abraham & Shaw, Part 2, p. 106.

Figure 6.2. Kupka, Amorpha, Fugue in Two Colors, 1912, #68 from the DMA catalog. (Andel, 1996, p. 129) , modified to accentuate the outlines suggesting Poincare's tangle.

CONCLUSION

Let us now take stock. We will begin with a review of the themes, and the mentalities and shifts of each chapter, and end with a reconsideration of our subtheme of math pedagogy.

Introduction

In the Introduction, we set out some goals and expectations for our six chapters. First are the three main themes:

• Dynamical historiography
• Math mentality
• Bolts from the blue

Then we expanded math mentality into our sequence of five cultural ecologies, each with its own math mentality, RGADX:

• Upper Paleolithic cultural ecology, aRithmetic Mentality (R)
• Riverine, Geometric Mentality (G)
• Islamic, Algebraic Mentality (A)
• Baroque, Galilean Dynamical Mentality (D)
• Modernist, Chaos Dynamical Mentality (X)

These are separated by four bifurcations, or paradigm shifts:

• R/G, 4000 BCE
• G/A, 800 CE
• A/D, 1600
• D/X, 2000

Finally, we explained our motivating subtheme: a modified program for teaching math in schools to avoid math anxiety.

After reading these six chapters, you have grasped our major themes as described in the Introduction. Mathematics is a large, integral part of the evolution of our collective consciousness and world cultural history, far larger and more exciting than the tiny fragment identified as mathematics by our current school program. Its development, much like the psychological development of a child, has proceeded in five stages: RGADX. These are connected by four cultural bifurcations, the fourth of which is in progress right now.

Chapter 1

In Chapter 1, on the Venus of Lespugue, we saw how arithmetic was expertly used by prehistoric musicians and sculptors. Art and mathematics were intimately connected to each other and everything else. This is the epitome of the aRithmetic Mentality of the late paleolithic time, and is also an exemplary bolt from the blue in which musicians brought forth an advanced form of arithmetic, musical arithmetic. R-people, natives of the aRithmetic Mentality, love to count, make lists, measure, and so on. The Venus of Lespugue is a measured object, displaying superb accuracy.

Chapter 2

In Chapter 2, on the Bethels of Scotland, we saw how solid geometry—including the most important shapes of ancient Greece—was explored, loved, and preserved by the megalith builders of prehistoric Britain, an outpost or precursor of the Riverine cultural ecology. The bethels are exemplary of the Geometric Mentality, and also a fabulous bolt of precognition of the Pythagorean solids. Inspired by the vault of the starry sky, extrapolated into a celestial sphere, these G-people, intellectual pioneers of the Geometric Mentality, explored the pure shapes of mathematical solids, first in clay perhaps, and then in the hardest stone. In doing so, they triggered the R/G shift.

Chapter 3

In Chapter 3, on the origins of algebra, we saw a mathematical explosion resulting from a reaction between different cultural styles.

Prehistoric (mental) arithmetic was given new life in the cultural ambiance of literacy and calligraphy of the Islamic renaissance. Again, this epitomizes the Algebraic Mentality. A-people love to compress a problem into a compact formula, then unpack it into an ornate solution. Early Islam was home to the G/A shift. In this context, we also revealed more of the conceptual frame of dynamical historiography: the reaction-diffusion metaphor of biological morphogenesis.

Chapter 4

In Chapter 4, on Fra Angelico, we learned of precognitive mathematical visions—of 20th century topology and geometry—on the part of Renaissance poets and painters, long before the development of these subjects by modern mathematicians. This sequence—poet, painter, mathematician—may be characteristic of the cultural shift process.

Chapter 5

D-people understand motion. In Chapter 5, on Vincenzo, we saw the clash of huge cultural forces surrounding the birth of a major new mathematical style, the Dynamical Mentality, and along with it, the birth of modern science. A preamble on Renaissance music theory provided the basis for our analysis of the coordinated shifts: A/D mentality shift, Renaissance to Baroque paradigm shift, and the birth of modern science. This sequence of shifts reveals the crucial role played by the arts and mathematics in the dynamics of cultural change. This was our first example of a bifurcation complex, a nested sequence of shifts.

Chapter 6

And in Chapter 6, on Poincaré, Kupka, and Satie, we see the computer graphics of the original computational mathematicians of the D/X shift, Ueda and Mandelbrot, rendered visible by an abstract painter, long before the invention of computers. The clairvoyance of Kupka, the first person to see the homoclinic tangle, and Satie, the first to hear a chaotic attractor, seems truly paranormal.

In sum

RGADX: Bolts from blue! We have discovered a bolt pattern for cultural bifurcations: artists, mathematicians, scientists.

How much of this would you recognize as mathematics on the basis of your school education? How much appears on standard tests? And what kind of training best supports the development of your natural mathematical skills? How is the integration of mathematics and culture faring today?

These six stories—in addition to being examples of cultural ecologies, math mentalities, bolts from the blue, and so on—exhibit the integration of math and cultural history, the basis of a new way of teaching math in school.

My goal here is to save mathematics, and our civilization in fact, from the dustbin. For innocent errors have crept into our school programs and become stuck there. Most people become deprived of their love of math, and develop math anxiety. Our cultural evolution has a broken bolt, and many world problems may follow from this epidemic disease. Math anxiety slows the species-wide advance in the understanding of complex systems, while the complexity of our world leaps ahead. We watch helplessly as world problems grow. The bugs in our education system must be corrected. A major paradigm shift is required in our basic school programs, in fact, to restore our proper trajectory of cultural evolution. And now that a major bifurcation is upon us, the D/X shift, we have a golden opportunity. The outcome is up to us. This book is intended as a contribution to this global shift.

BIBLIOGRAPHY

Abbott, Edwin Abbott. *Flatland; a Romance of Many Dimensions, with illus. by the author; a square.* New York, Barnes & Noble, 1963.

Abraham, Ralph. *Chaos, Gaia, Eros.* San Francisco: Harper-Collins, 1994.

Abraham, Ralph. *Bifurcation of the Kalihari !Kung, J.* World Futures, 1995.

Abraham, Ralph. *Chaos and the millennium,* Alexandria 5, 2001.

Abraham, Ralph, Terence McKenna, and Rupert Sheldrake. *Trialogues at the Edge of the West.* Santa Fe, NM: Bear, 1992.

Abraham, Ralph, Terence McKenna, and Rupert Sheldrake. *Evolutionary Mind: Trialogues at the Edge of the Unthinkable.* Santa Cruz, CA: Aerial Press, 1998.

Abraham, Ralph, Terence McKenna, and Rupert Sheldrake. *Chaos, Creativity, and Cosmic Consciousness.* Rochester, VT: Park Street Press, 2001.

Abraham, Ralph, and Christopher Shaw. *Dynamics, the Geometry of Behavior, 3rd edn.* Aerial Press, 1999.

Ainsley, Robert, ed. *The Encyclopedia of Classical Music.* London: Carlton Books, 1999.

Andel, Jaroslav, and Dorothy Kosinski, eds. *Painting the Universe: Frantisek Kupka, Pioneer in Abstraction,* Dallas: Dallas Museum of Art, 1997.

Apostle, Hippocrates G. *Aristotle's Metaphysics.* Bloomington, IN: Indiana University Press, 1966.

Aristotle. *On Man in the Universe: Metaphysics, Parts of Animals, Ethics, Politics, Poetics.* Roslyn, NY: Walter J. Black, 1943.

Backhouse, Janet. *The Lindisfarne Gospels,* Oxford: Phaidon, 1981.

Baker, Arthur. *Celtic Hand, Stoke by Stroke: Irish Half-Uncial from "The Book of Kelis".* New York: Dover, 1983.

Baines, John C., and Cormac O Cuilleanain. *Dante and the Middle Ages: Literary and Historical Essays,* Dublin: Irish Academic Press, 1995.

Batish, Shiv Dayal, and Ashwin Batish. *Ragopedia, Volume One: Exotic Scales of North India*, Santa Cruz, CA: Batish Publications, 1989.

Bergin, Thomas G. *Dante*, New York: Orion Press, 1965.

Bemal, Martin. *Black Athena: the Ajroasiatic Roots of Classical Civilization*, 2 vols., New Brunswick, N..l.: Rutgers University Press, 1987, 1991.

Bent, Ian. *New Grove Dictionary of Music and Musicians*. Basingstoke: Macmillan, 1987.

Bochner, Salomon. *The Role of Mathematics in the Rise of Science*. Princeton: Princeton University Press, 1966.

Bourke, Vemon J. *The Pocket Aquinas*. New York: Pocket Books, 1960.

Bowie, Theodore, ed. *The Sketchbook of Villard de Honnecourt, 2nd edn.*, Bloomington, IN: Indiana University Press, 1959.

Boyer, Carl B. *A History of Mathematics*, Princeton: Princeton University Press, 1968.

Browning, Robert. Teachers, in: Gulielmo Cavallo, ed., *The Byzantines*. Chicago: University of Chicago Press, 1997; pp. 95-116.

Brumbaugh, Robert S. *Plato's Mathematical Imagination: The Mathematical Passages in the Dialogues and Their Interpretation*. Bloomington: Indiana University Press, 1954/1964.

Bukofzer, Manfred E. *Music in the Baroque Era – From Monteverdi to Bach*. New York: Norton, 1947.

Bungener, L. F. *History of the Council of Trent, tr. David D. Scott*. New York: Harper, 1855.

Burckhardt, Jacob. *The Age of Constantine the Great, Moses Hadas, tr.* Garden City, NY: Doubleday, 1852/1949.

Burckhardt, Jacob *The Civilization of the Renaissance in Italy, S. G. C. Middlemore, tr.* London: Phaidon Press, 1860/1944.

Cajori, Florian. *A History of Mathematical Notations, La Salle, lll.* Open Court Pub. Co., 1928/ 1951.

Canfora, Luciano. *The Vanished Library: A Wonder of the Ancient World.* Berkeley, CA: University of California Press, 1987/ 1990.

Carpenter, Edmund, and Carl Schuster. *Patterns that Connectional Symbolism in Ancient and Tribal Art*. New York: Harry N. Abrams, 1996.

Chadwick, Henry. *The Early Church*. New York: Dorset Press, 1967/ 1986.

Chauvet, Jean-Marie, Eliette Brunel Deschamps, and Christian Hill-aire. *Dawn of Art: the Chauvet Cave, the Oldest Known Paintings in the World*, New York: Henry Abrams, 1996.

Cho, Gene J. *The Discovery of Musical Equal Temperament in China and Europe in the Sixteenth Century*. Lewiston, NY: E. Mellen Press, 2003.

Cohen, H. F. *Quantifying Music: the Science of Music at the First Stage of the Scientific Revolution, 1580-1650*. Boston: Reidel, 1984.

Copenhaver, Brian P. *Hermetica: the Greek Corpus Hermeticum and the Latin Asclepius in a New English Translation with Notes and Intro-duction*. Cambridge: Cambridge University Press, 1992.

Cornford, F. M. *Before and After Socrates*. Cambridge: Cambridge University Press, 1932/ 1992.

Cotterell, Arthur. *The Penguin Encyclopedia of Ancient Civilizations*. London: Penguin Books, 1980.

Coulton, G. G. *Inquisition and Liberty*. Boston: Beacon Press, 1938/1959.

Critchlow, Keith. *Time Stands Still: New Light on Megalithic Science*. London: Gordon Fraser, 1979.

Crosby, Alfred W. *The Measure of Reality: Quantnication and West-ern Society, 1250-1600,* Cambridge: Cambridge University Press, 1997.

Cross, Richard. *Duns Scotus*. New York; Oxford: Oxford University Press, 1999.

Damasio, Antonio. *Descartes' Error: Emotion and Human Reason*, New York: Basic Books, 1994, also his Damasio, Antonio, *The Feeling of What Happens: Body and Emotion in the Making of Consciousness*, New York: Harcourt Brace, 1999.

Davies, Brian. *The Thought of Thomas Aquinas*. Oxford: Clarendon Press, 1992.

Dehaene, Stanislas. *The Number Sense: How the Mind Creates Math-ematics*. New York: Oxford University Press, 1997.

de Nicolas, Antonio T. *Four-Dimensional Man: The Philosophical Methodology of the Rg Veda*. Bangalore: Dharmaram College, 1971.

de Nicolas, Antonio T. *Meditations Through the Rg Veda*. York Beach, ME: Nicolas Hays, 1976.

de Nicolas, Antonio T. *Avatara: The Humanization of Philosophy Through the Bhagavad Gita*, York Beach, ME: Nicolas Hays, 1976.

de Santillana, Giorgio, *The Crime of Galileo*. New York: Time, Inc., 1955/1962.

de Santillana, Giorgio, and Hertha von Dechend. *Hamlet's Mill*. Boston: Gambit, 1969.

Diehl, Charles, *Byzantium: Greatness and Decline*, Naomi Walford, tr., Peter Charanis, ed. New Brunswick, NJ: Rutgers University Press, 1919/1957.

Dieudonne, Jean Alexandre. *A History of Algebraic and Differential Topology, 1900-1960*. Boston: Birkhauser, 1989.

Dijksterhuis, Eduard Jan. *The Mechanization of the World Picture*. Oxford: Clarendon Press, 1961.

Diodorus Siculus, Books II, 35 -IM 58, Trans. C. H. Oldfather. Harvard Univ. Press, Loeb Classical Library: Cambridge, MA, 1935.

Diringer, David. *The Alphabet, A Key to the History of Mankind, 3rd edn*. New York: Funk & Wagnalls, 1968.

Donald, Merlin. *Origins of the Modern Mind: Three Stages in the Evolution of Culture and Cognition*. Cambridge, Mass.: Harvard University Press, 1991.

Drake, Stillman. *Galileo*. New York: Hill and Wang, 1980.

Drake, Stillman. *Essays on Galileo and the History and Philosophy of Science, v. 1*. Toronto: University of Toronto Press, 1995.

Eisler, Riane. *The Chalice and the Blade: Our History, Our Future*. Cambridge, MA: Harper & Row, 1987.

Farmer, Steve, John B. Henderson, and Michael Witzel. *Neurobiology, layered texts, and correlative cosmologies: A cross-cultural framework for premodern history, Bulletin of the Museum of Far Eastern Antiquities*. Stockholm, Sweden, September, 2002, pp. 1-11.

Feldhay, Rivka. *Galileo and the Church: Political Inquisition or Critical Dialogue?* Cambridge: Cambridge University Press, 1995.

Fideler, David. A note on Ptolemy's polychord and the contemporary relevance of harmonic science. *Alexandria* 2, 1993, pp. 167-182.

Fideler, David. *Platonic Academies*. Preprint.

Field, Arthur. *Origins of the Platonic Academy of Florence*. Princeton, NJ: Princeton University Press, 1988.

Fleck, Ludwik. *Genesis and Development of a Scienttjic Fact; translated by Fred Bradley and Thaddeus J. Trenn*. Chicago: University of Chicago Press, 1979.

Fleming, William. *Arts and Ideas, 6th ed.* New York: Holt, Rinehart and Winston, 1955/1980.

Forster, E. M., *Alexandria, A History and a Guide*. Gloucester, MA: Peter Smith, 1968.

Fowden, Garth. *The Egyptian Hermes: a Historical Approach to the Late Pagan Mind*. Princeton, NJ.: Princeton University Press, 1986/ 1993.

Fox, Matthew, and Rupert Sheldrake. *The Physics of Angels: Exploring the Realm where Science and Spirit Meet*. San Francisco: Harper San Francisco, 1996.

Freedberg, David. *The Eye of the Lynx: Galileo, his Friends, and the Beginnings of Modern NaturalHistory*. Chicago: University of Chicago Press, 2002.

Friedlaender, Walter. *Mannerism and Anti-mannerism in Italian Painting*. New York: Columbia University Press, 1957/1990.

Galilei, Vincenzo. *Dialogue on Ancient and Modern Music, tr C. Palisca*. New Haven: Yale University Press, 1581/2003.

Gamble, Clive. *Interaction and alliance in paleolithic society,* Man, 17, 1982, pp. 92-107.

Gebser, Jean. *The Ever-present Origin. Transl. by Noel Barstad and Algis Mickunas*. Athens, Ohio: Ohio University Press, 1984, c1985.

Geneva, Ann. *Astrology and the Seventeenth Century Mind: William Lilly and the Language of the Stars*. Manchester: Manchester University Press, 1995.

Gibb, H. A. R. *Mohammedanism*. New York: Oxford University Press, 1962.

Gillispie, Charles, ed. *Dictionary of Scientyic Biography*. New York: Scribner, 1970.

Gimbutas, Marija Alseikaite. *The Language of the Goddess: Unearthing the Hidden Symbols of Western Civilization*. San Francisco: Harper & Row, 1989.

Grant, Edward. *Physical Science in the Middle Ages.* New York: Wiley, 1971.

Graziosi, Paolo. *Paieolifhie Ari.* New York: McGraw-Hill, 1960.

Godwin, Joscelyn. *Harmonies of Heaven and Earth: The Spiritual Dimensions of Music.* Rochester, VT: Inner Traditions, 1987.

Godwin, Joscelyn, ed. *Cosmic Music, Musical Keys to the Interpretation of Reality: Essays by Marius Schneider: Rudolf Haase, Hans Erhard Lauer.* Rochester, VT: Inner Traditions, 1989.

Guthrie, Kenneth Sylvan. *The Pythagorean Sourcebook and Library: an Anthology of Ancient Writings which Relate to Pythagoras and Pythagorean Philosophy.* Grand Rapids, MI: Phanes Press, 1920/ 1987

Guthrie, Kenneth Sylvan. *Porphyry's Launching Points to the Realm of the Mind: An Introduction to the Neoplatonic Philosophy of Plotinus.* Grand Rapids, MI: Phanes Press, 1988.

Guthrie, W. K. C. *Orpheus and Greek Religion: a Study of the Orphic Movement.* Princeton, NJ: Princeton University Press, 1952/1993.

Harding, James. *Erik Satie.* London: Seeker and Warburg, 1975.

Havelock, Eric. *Preface to Plato.* Cambridge, UK: Cambridge University Press, 1991.

Hawkins, Gerald S. *Beyond Stonehenge.* New York: Harper & Row, 1973.

Healey, John F. *The Early Alphabet.* Berkeley: University of California, 1990.

Heath, Thomas L. *Euclid, the Thirteen Books of the Elements, 2nd ed.* New York: Dover, 1908/ 1960.

Heath, Thomas L. *A History of Greek Mathematics; Volume 1, From Thales to Euclid.* New York: Dover, 1921/1956.

Henderson, Linda Dalrymple. *X Rays and the quest for invisible reality in the art of Kupka, Duchamp, and the Cubists.* Art Journal, Winter 1988, pp. 323-340.

Henderson, Linda Dalrymple. *The Fourth Dimension and Non-Euclidean Geometry in Modern Art.* Princeton: Princeton University Press, 1983.

Henderson, Linda Dalrymple. *Duchamp in Context: Science and Technology in the "Large Glass" and Related Works.* Princeton: Princeton University Press, 1998.

Joseph, George Gheverghese. *The Crest of the Peacock, Non-European Roots of Mathematics,* London: Penguin, 1992.

Jowett, Benjamin. *The Dialogues of Plato, 3rd ed.* New York: Random House, 1892/ 1937.

Katz, Victor. *A History of Mathematics: an Introduction.* New York: HarperCollins, 1993.

Kayser, Hans. *Ein Harmonikaler Teilungskanon,* Basel: Schwabe, 1946.

Khatibi, Abdelkebir, and Mohammed Sijelmassi. *The Splendour of Islamic Calligraphy.* London: Thames and Hudson, 1996.

Klein, Jacob. *Greek Mathematical Thought and the Origin of Algebra.* New York: Dover, 1968/1992.

Kosinski, Dorothy, and Jaroslav Andel, eds. *Frantisek Kupka: die Abstracten Farben des Universums.* Ostfeldern bei Stuttgart: Verlag Gerd Hatje, 1998.

Koyré, Alexandre. *The Astronomical Revolution; Copernicus, Kepler; Borelli.* Paris, Hermann, 1961. Ithaca, N.Y.: Cornell University Press, 1973.

Kramer, Samuel Noah. *History Begins in Sumer: Thirty-nine Firsts in Man's Recorded History.* Garden City, NY: Doubleday, 1959; Philadelphia: University of Pennsylvania Press, 1981.

Kristeller, Paul Oskar. *The Philosophy of Marsilio Ficino; transl. by Virginia Conant.* Gloucester, MA: P. Smith, 1943/ 1964.

Kuhn, Thomas. *The Structure of Scientnic Revolutions.* Chicago: University of Chicago Press, 1962.

Lakoff, George, and Rafael E. Nunez. *Where Mathematics Comes From: How the Embodied Mind brings Mathematics into Being.* New York: Basic Books, 2000.

Lane-Poole, Stanley. *The Story of the Moors in Spain.* Baltimore, MD: Black Classic Press, 1886/1990.

Leff, Gordon. *William of Ockham: The Metamorphosis of Scholastic Discourse.* Manchester: Manchester University Press, 1975.

Le Goff, Jacques. *Intellectuals in the Middle Ages, Teresa Lavender Fagan, tr.* Oxford: Blackwell, 1957/ 1993.

Leroi-Gourhan, Andre. *Treasures of Prehistoric Art.* New York: H. N. Abrams, 1967.

Leroi-Gourhan, *Andre, The Art of Prehistoric Man in Western Europe*. London: Thames and Hudson, 1968.

Levin, Flora R. *The Manual of Harmonics of Nicomachus the Pythagorean*. Grand Rapids, MI: Phanes, 1994.

Lewis, Archibald. *Nomads and Crusaders: 1000-1368*. Bloomington, ID: University of Indiana Press, 1988.

Lewis, Bernard. *What Went Wrong: Western Impact and Middle Eastern Response*. New York: Oxford University Press, 2002.

Lewis-Williams, David. *The Mind in the Cave: Consciousness and the Origins of Art*. London: Thames & Hudson, 2002.

Lloyd, Christopher, *Fra Angelico*. London: Phaidon Press, 1979.

Lloyd, G. E. R. *Early Greek Science: Thales to Aristotle*. London: Chatto & Windus, 1970.

Lloyd, G. E. R. *Greek Science after Aristotle*. New York: Norton, 1973.

Lovelock, James. *Gaia New Look at Life on Earth*. Oxford: Oxford University Press, 1995.

Mandelbrot, Benoit B. *Fractals: Form, Chance, and Dimension*. San Francisco: W. H. Freeman, 1977.

Mandelbrot, Benoit B. *The Fractal Geometry of Nature*. San Francisco: W.H. Freeman, 1982.

Marshack, Alexander. *The Roots of Civilization: the Cognitive Beginnings of Man's First Art, Symbol, and Notation*. New York: McGraw-Hill, 1972.

Marshack, Alexander. *Ice Age Art*. New York: American Museum of Natural History, 1979.

Marshack, Alexander. *Hierarchical Evolution of the Human Capacity: The Paleolithic Evidence*. New York: American Museum of Natural History, 1985.

Marshall, Dorothy N. *Carved stone balls*. Proc. Soc. Antiquaries of Scotland, v. 108 (1976-77), pp. 40-72.

Marshall, Dorothy N. *Further notes on carved stone balls*.Proc. Soc. Antiquaries of Scotland, v. 1 13 (1983), pp. 62.8-630.

McClain, Emest G. *The Myth of Invariance: The Origin of the Gods, Mathematics and Music from the Rg Veda to Plato*. Boulder, CO: Shambala, 1976/ 1978.

McClain, Ernest G. *The Pythagorean Plato: Prelude to the Song Itself.* York Beach, ME: Nicolas-Hays, 1978.

McClain, Emest G. *Meditations though the Quran: Tonal Images in an Oral Culture.* York Beach, ME: Nicolas-Hays, 1981.

McClain, Ernest G. *Tonal isomorphism in Plato and the I Ching: Brumbaugh as cultural anthropologist. In: Plato, Ume, and Education: Essays in Honor of Robert S. Brumbaugh.* Brian P. Hendley, ed. Albany NY: State University of New York Press, 1987.

McLuhan, Marshall, and Eric McLuhan. *Laws of Media: the New Science.* Toronto: University of Toronto Press, 1988.

Mead, G. R. *Thrice-greatest Hermes: Studies in Hellenistic Theosophy and Gnosis.* London, 1906. New York, 1992.

Meehan, Bemard. *The Book of Kells: An Illustrated Introduction. to the Manuscript in Trinity College Dublin.* London: Thames and Hudson, 1994.

Menninger, Karl. *Number Words and Number Symbols, A Cultural History of Numbers.* New York: Dover, 1969, 1992.

Miller, Arthur I. *Einstein and Picasso: Space, Time, and the Beauty that Causes Havoc.* New York: Basic Books, 2001.

Morichiello, Paulo. *Fra Angelico: The San Marco Frescoes.*London: Thames and Hudson, 1996.

Morrow, Glenn R.,tr., Proclus. *A Commentary on the First Book of Euclid's Elements.* Princeton, NJ: Princeton University Press, 1970/ 1992.

Mountford, J. F. *Greek music and its relation to modem times.* J. Hellenic Studies 40, 1920. pp. 13-42.

Mueller, Ian. *Philosophy of Mathematics and Deductive Structure in Euclid's Elements.* Cambridge, MA: MIT Press, 1981.

Murray, Peter and Linda. *The Art of the Renaissance.* London: Thames and Hudson, 1963.

Myers, Rollo H. *Erik Satie.* St. Clair Shores, Michigan: Scholarly Press, 1947/1977.

Naveh, Joseph. *Early History ofthe Alphabet: An Introduction to West Semitic Epigraphy and Paleography.* Jerusalem: Magnes Press, 1982.

Neugebauer, Otto. *The Exact Sciences in Antiquity, 2d ed.* New York: Harper, 1962.

Nigosian, Solomon. *Islam: the Way of Submission.* Welling-borough: Aquarian Press, 1962.

O'Brien, Elmer, tr. *The Essential Plotinus.* Indianapolis, IN: Hackett, 1964/1975.

Palacios, Miguel Asin. *Islam and the Divine Comedy.* London: Frank Cass, 1968.

Palisca, Claude V. *Baroque Music.* Engelwood Cliffs, NJ: Prentice-Hall, 1968.

Palisca, Claude V. *The Florentine Camerata.* New Haven: Yale University Press, 1989.

Parsons, Edward Alexander. *The Alexandrian Library, Glory of the Hellenic World; its Rise, Antiquities, and Destructions.* Amsterdam, NY: Elsevier Press, 1952.

Peterson, Mark A. *Dante and the 3-sphere*, Am. J. Phys. 47(12), Dec. 1979, PP. 1031-1035.

Pope-Hennessy, John. *Fra Angelico.* London: Phaidon Press, 1952.

Pope-Hennessy, John. *Angelico.* Firenze: Scala Books, 1981

Purse, Jill. *Spirals.* London: Thames and Hudson, 1974.

Reese, Gustave. *Music in the Renaissance.* New York: Norton, 1954/59.

Rey, H. A. *The Stars: A New Way to See Them.* Boston: Houghton Mifflin, 1952/1970.

Ridondi, Pietro. *Galileo Heretic; transl. by Raymond Rosenthal.* Princeton: Princeton University Press, 1987.

Rogers, Elizabeth Frances. *Peter Lombard and the Sacramento/ System.* New York: 1917.

Rosen, Frederic, tr. *Al-Khwarazmfs Algebra.* Islamabad: Pakistan Hira Council, 1831/ 1989.

Sayili, Aydin, Introduction, in Rosen, 1989.

Schimmel, Annemarie. *Calligraphy and Islamic Culture.* New York: New York University Press, 1984.

Schlesinger, Kathleen. *The Greek Aulos; a Study of its Mechanism and of its Relation to the Modal System of Ancient Greek Music, Followed by a Survey of the Greek Harmoniai in Survival or Rebirth in Folk-Music.* London: Methuen, 1939.

Schmandt-Besserat, Denise. *How Writing Came About*. Austin, TX: University of Texas Press, 1996.

Schneider, Marius. *Cosmic Music: Musical Keys to the Interpretation of Reality.* Rochester, Vt.: Inner Traditions, 1989.

Scholem, Gershom G. *Major Trends in Jewish Mysticism.* New York: Schoken, 1941/ 1946.

Schultz, Joachim. *Movement and Rhythms of the Stars: A Guide to Naked-eye Observation of Sun, Moon, and Planets.* Edinburgh: Floris Books, 1963/ 1986.

Shannon, Albert Clement. *The Medieval Inquisition.* Washington: Augustinian College Press, 1983.

Shapiro, Herman. *Motion, Time and Place According to William Ockham.* St. Bonaventure, NY: The Franciscan Institute, 1957.

Shlain, Leonard. *Art and Physics: Parallel Visions in Space, Time, and Light.* New York: Morrow, 1991.

Shlain, Leonard. *The Alphabet versus the Goddess: the Conflict between Word and Image.* New York: Viking, 1998.

Shultz, Hellmuth. *Callanish, a Guide to the Standing Stones and the Callanish Complex, 3rd ed.* Callanish: B. M. Schultz, 1983/1990.

Slonimsky, Nicolas. *The Portable Baker's Biographical Dictionary of Musicians.* New York: Schirmer, 1995.

Smale, Stephen. *The Mathematics of Yime: Essays on Dynamical Systems, Economic Processes, and Related Topics.* New York: Springer-Verlag, 1980.

Smith, Herbert W. *The Greatness of Dante Alighieri.* Bath: Bath University Press, 1973.

Snyder, Tango Parish. *The Biosphere Catalogue.* London: Synergetic Press, 1985.

Soffer, Olga, Pamela Vandiver, Martin Olivia, and Ludik Seitel. *The case of the exploding goddess figurines.* Archeology, January/February 1993.

Spade, Paul Vincent, ed. *The Cambridge Companion to Ockham.* Cambridge: Cambridge University Press, 1999.

Spike, John T. *Fra Angelico.* New York: Abbeville Press, 1997.

Stanley, Thomas, tr. *The Chaldean Oracles as Set Down by Julianus.* Gilette. NJ: Heptangle, 1989.

Templier, Pierre-Daniel. *Erik Satie.* Cambridge, MA.: MIT Press, 1969.

Teipstra, Siemen. *An introduction to the monochord.* Alexandria 2, 1993, PP. 137-166.

Thom, Alexander. Megalithic Lunar Observatories. Oxford: Clarendon Press, 1971.

Thompson, William Irwin. *The Time Falling Bodies Take to Light: Mythology, Sexuality, and the Origins of Culture.* New York: St. Martin's Press, 1981.

Thompson, William Irwin. *Pacific Shift.* San Francisco: Sierra Club Books, 1985.

Thompson, William Irwin. *Imaginary Landscape: Making Worlds of Myth and Science.* New York: St. Martin's Press, 1989.

Thompson, William Irwin. *Coming into Being: Artifacts and Texts in the Evolution of Consciousness.* New York: St. Martin's Press, 1996.

Thompson, William Irwin. *Transforming History: A Curriculum for Cultural Evolution.* Great Barrington, MA: Lindisfarne Books, 2001.

Thompson, William Irwin. *Literary-mathematical mentalities and the evolution of culture, Journal of Consciousness Studies.* January 2004.

Thompson, William Irwin. *Self and Society: Studies in the Evolution of Culture.* London: Imprint Academic, 2004.

Thompson, William Irwin. *Transforming History: A New Curriculum for a Planetary Culture.* Great Barrington, MA: Lindisfarne Books, 2008.

Thompson, William Irwin. *Self and Society (Societas).* London: Imprint Academic, 2009.

Todd, Nancy Jack, and John Todd. *From Eco-Cities to Living Machines: Principles of Ecological Design.* Berkeley, CA: North Atlantic Books, 1994.

Turner, Howard R. *Science in Medieval Islam: an Illustrated Introduction.* Austin: University of Texas Press, 1997.

Varela, Francisco, ed. *Sleeping, Dreaming, and Dying: A Conversation with the Dalai Lama*, ed. Boston: Wisdom Publications, 1997.

Waddington, Conrad. *Behind Appearance: A Study of the Relations between Painting and the Natural Sciences in this Century.* Edinburgh: Edinburgh University Press, 1969. Cambridge, MA: MIT Press, 1970.

Walker, C. B. F. *Cuneiform*, London: The British Museum, 1987.

Walker, D. P. *Spiritual and Demonic Magic from Ficino to Campanella.* London: Warburg Institute, 1958.

Werner, Eric. *The Sacred Bridge.* New York: Columbia University Press, 1959.

Woodhouse, C. M. *George Gemistos Plethon: the Last of the Hellenes.* Oxford: Clarendon Press, 1986.

Zarlino, Gioseffo (1558/1968). *The Art of Counterpoint.* tr. Guy A. Marco and Claude V. Palisca. New York: Norton, 1558/1968.

Zhang, Juzhong, Garman Harbottle, Changsui Wang, and Zhaochen Kong. Oldest playable musical instrument found at Jiahu early Neolithic site in China, *Nature* 401:366-368, 1999.

INDEX

A

Algebraic Mentality 7, 51, 79, 80, 105, 107
archeoastronomy 11, 41
archeomusicology 11
Archytas 42
Aristotle 44, 109, 116
Aristoxenos 20
aRithmetic Mentality 7, 11, 33, 105, 106
Atlantis 43

B

Babylonia 13, 35, 39, 41, 59
Benoit Mandelbrot 8, 95, 98
bethels 7, 33, 37, 38, 41, 42, 50, 106, 123
Bethels of Scotland xv, 7, 33, 106
bifurcations 1, 2, 3, 4, 5, 55, 79, 101, 105, 106, 108
bolt from the blue xiv, 6, 7, 8, 33, 60, 65, 82, 95, 101, 106
Bolts fom the blue 1

C

Callanish 34, 35, 37, 38, 39, 50, 119
Calligraphy 51
Camerata 88
canon 16
Canon of Lespugue 11, 21
chaos dynamical mentality 2, 7
Chaos, Gaia, Eros ii, 2, 9, 55, 109
chaos theory 1, 2, 8, 9, 11, 95, 97, 99, 101, 102
chaotic dynamical system 96
Cipriano de Rore 88
complex dynamical system 1, 2, 5, 53
construction 40, 42, 45, 46, 66, 72
cosmic figures 7, 33, 38, 39, 40, 41, 42, 43, 44, 47, 48
cultural ecology 1, 3, 4, 5, 6, 7, 11, 46, 51, 105, 106

D

Davidic set 19
delay convention 79
diatonic scale 19, 43
Diodorus Siculus 34, 40, 50, 112
dissonance 88